AF756435

Travel Behaviour

TRANSPORT ECONOMICS, MANAGEMENT AND POLICY

General Editor: Kenneth Button, *Professor of Public Policy, School of Public Policy, George Mason University, USA*

Transport is a critical input for economic development and for optimizing social and political interaction. Recent years have seen significant new developments in the way that transport is perceived by private industry and governments, and in the way academics look at it.

The aim of this series is to provide original material and up-to-date syntheses of the state of modern transport analysis. The coverage embraces all conventional modes of transport but also includes contributions from important related fields such as urban and regional planning and telecommunications where they interface with transport. The books draw from many disciplines and some cross disciplinary boundaries. They are concerned with economics, planning, sociology, geography, management science, psychology and public policy. They are intended to help improve the understanding of transport, the policy needs of the most economically advanced countries and the problems of resource-poor developing economies. The authors come from around the world and will represent some of the outstanding young scholars as well as established names.

Titles in the series include:

Air Transport Networks
Theory and Policy Implications
Kenneth Button and Roger Stough

Analytical Transport Economics
An International Perspective
Edited by Jacob B. Polak and Arnold Heertje

Intelligent Transport Systems
Cases and Policies
Edited by Roger R. Stough

Transport and Environment
In Search of Sustainable Solutions
Edited by Eran Feitelson and Erik T. Verhoef

Environmental Costs and Liberalization in European Air Transport
A Welfare Economic Analysis
Youdi Schipper

Reforming Transport Pricing in the European Union
A Modelling Approach
Edited by Bruno De Borger and Stef Proost

Travel Behaviour
Spatial Patterns, Congestion and Modelling
Edited by Eliahu Stern, Ilan Salomon and Piet H.L. Bovy

Travel Behaviour

Spatial Patterns, Congestion and Modelling

Edited by

Eliahu Stern

Professor of Geography and Environmental Development, Ben Gurion University of the Negev, Israel,

Ilan Salomon

Professor of Geography and Transport Planning, Department of Geography, Hebrew University of Jerusalem, Israel

Piet H.L. Bovy

Professor of Transport Planning, Delft University of Technology, The Netherlands

TRANSPORT ECONOMICS, MANAGEMENT AND POLICY SERIES

Edward Elgar

Cheltenham, UK • Northampton, MA, USA

388.31409(4)
T

© Eliahu Stern, Ilan Salomon and Piet H.L. Bovy 2002

All rights reserved. No part of this publication may be reproduced, stored in a retrieval system or transmitted in any form or by any means, electronic, mechanical or photocopying, recording, or otherwise without the prior permission of the publisher.

Published by
Edward Elgar Publishing Limited
Glensanda House
Montpellier Parade
Cheltenham
Glos GL50 1UA
UK

Edward Elgar Publishing, Inc.
136 West Street
Suite 202
Northampton
Massachusetts 01060
USA

CARDIFF UNIVERSITY
-2 JUL 2002
PRIFYSGOL CAERDYDD

A catalogue record for this book
is available from the British Library

Library of Congress Cataloguing in Publication Data
Travel behaviour : spatial patterns, congestion and modelling / edited by Eliahu Stern, Ilan Salomon, Piet H.L. Bovy.
p. cm – (Transport economics, management, and policy)
Includes bibliographical references and index.
1. Traffic flow–Europe. 2. Telecommunication–Europe. 3. Transportation–Europe. I. Stern, Eliahu. II. Salomon, Ilan. III. Bovy, Piet H.L., 1943- IV. Series.

HE336.T7 T735 2002
388.3'14'094–dc21

2001040971

ISBN 1 84064 707 8

Typeset by Manton Typesetters, Louth, Lincolnshire, UK.
Printed and bound in Great Britain by MPG Books Ltd, Bodmin, Cornwall.

To NECTAR

Network of European Communications and Transportation Activity Research

Contents

List of figures

List of tables

List of contributors

Kay W. Axhausen
IVT, ETH–Hoenggerberg, CH–8093 Zurich, Switzerland.
Tl. 0041-1-633 3943, Fax 0041-1-633 1057, Email: axhausen@ivt.baum.ethz.ch

Michael Bader
Ingenieurburp Koll, Agenbachsiedlung 16a, A-6060 Ampass, Germany.
Tl. +43 512 364 2750, Fax +43 512 364 2755, Email: verkhersplanung-badr@tirol.com

Piet H.L. Bovy
Transportation Planning and Traffic Engineering Section, Faculty of Civil Engineering and Geosciences, Delft University of Technology, 2628 CN Delft, The Netherlands.
Tl. +31 15 278 4611, Fax +31 15 278 3179, Email: bovy@ct.tudelft.nl

Alessandro Cento
MeeoP Institute, Faculty of Economics, Universita della Svizzera Italiana, Via Ospedale 13, CH–6900 Lugano, Switzerland.
Tl. +41-91-912-4731, Fax +41-91-912-4647, Email: centoa@lu.unisi.ch

Sean T. Doherty
Department of Geography and Environmental Studies, Wilfrid Laurier University, Waterloo, Ontario, Canada N2L 3C5.
Tl. 519 884 0710 (X2044), Fax 519 725 1342, Email: sdoherty@wlu-ca

Tommy Gärling
Department of Psychology, Göteborg University, P.O. Box 500, SE-40530 Goteborg, Sweden.
Tl. +46 0 31 773 1881, Fax +46 0 31 773 4628, Email: Tommy.Garling@mangrove.psy.gu.se

Hans Hilbers
TNO–Inro, Schoemakerstraat 97, P.O. Box 6041, 2600 JA Delft, The Netherlands.
Tl. +31 15 269 6878, Fax +31 15 269 7782, Email: hhi@inro.tno.nl

Veli Himanen
VTT Communication and Infrastructure, P.O. Box 1901, FIN 02044 VTT, Finland.
Tl. +358 9 456 4598, Fax +358 9 464 850, Email: veli.himanen@vtt.fi

Viviane Illegems
Business Economics and Strategic Management, Free University Brussels, Pleinlaan 2, 1050 Brussels, Belgium.
TL. +32 2 629 2129, Fax +32 2 629 2060, Email: Viviane.Illegems@vub.ac.be

Gysbertus R.M. Jansen
TNO–Inro, Schoemakerstraat 97, P.O.Box 6041, 2600 JA Delft, The Netherlands.
Tl. +31 15 269 6878, Fax +31 15 269 7782, Email: g.jansen@inro.tno.nl

Helmut Köll
Ingenieurburp Koll, Agenbachsiedlung 16a, A-6060 Ampass, Germany.
Tl. +43 512 364 2750, Fax +43 512 364 2755, Email: verkhersplanung-koel@tirol.com

Mervi Lehto
VTT Building Technology, P.O. Box 1804, FIN 02044 VTT, Finland.
Tl. +358 9 456 4570, Fax +358 9 456 4709, Email: mervi.lehto@vtt.fi

Kees Maat
OTB Research Institute for Housing, Urban and Mobility Studies, Delft University of Technology, P.O. Box 5030, 2600 GA Delft, The Netherlands.
Tl. +31 15 278 7640, Fax +31 15 278 3450, Email: maat@otb.tudelft.nl

Rico Maggi
MeeoP Institute, Faculty of Economics, Universita della Svizzera Italiana, Via Ospedale 13, CH–6900 Lugano, Switzerland.
Tl. +41 91 912 4636, Fax +41 91 912 4647, Email: Rico.Maggi@lu.unisi.ch

Eric J. Miller
Institute of Transportation, Traffic, Highways and Railways Engineering, Swiss Federal Institute of Technology, ETH–Hoenggerberg, CH–8093 Zurich, Switzerland.
Tl. +43 1 504 1258, Fax +43 1 504 3536.

Jean-Pierre Orfeuil
INRETS, 94114 Arcueil Cedex, France.
Tl. +33 1 474 0 7275, Fax +33 1 4547 5606, Email: orfeuil@inrets.fr

Ilan Salomon
Department of Geography, The Hebrew University, Jerusalem 91905, Israel.
Tl. +972 2 588 3345, Fax +972 2 582 0459, Email: msilans@huij.mscc.ac.il

Yoram Shiftan
Transportation Research Institute, Technion, Technion City, Haifa 32000, Israel.
Tl. +972 4 829 2381, Fax +972 4 822 5716, Email: yxs@tx.technion.ac.il

Rosette S'Jegers
Center for Business Economics and Strategic Management, Solvay Business School, Free University of Brussels, Pleinlaan 2, 1050 Brussels, Belgium.
TL. +32 2 629 2129, Fax +32 2 629 2060, Email: rosette.S'jegers@vub.ac.be

Eliahu Stern
Department of Geography & Environmental Development, Ben Gurion University of the Negev, Beer Sheva 84105, Israel.
Tl. +972 8 647 2005, Fax +972 8 647 2821, Email: elistern@bgumail.bgu.ac.il

Toon Van Der Hoorn
Transportation Research Center, Ministry of Transport, Rijkswaterstaat, 3000 BA Rotterdam, The Netherlands. Tl. +31 10 282 5746, Fax. +31 10 282 5642, Email: a.i.j.m.vdhoorn@avv.rws.minvenw.nl

Bert Van Wee
National Institute of Public Health and the Environment (RIVM), P.O. Box 1, 3720 BA Bilthoven, The Netherlands. Tl. +31 30 274 3654, Fax +31 30 274 4417, Email: bert.van.wee@rivm.nl

Alain Verbeke
Center for Business Economics and Strategic Management, Solvay Business School, Free University of Brussels, Pleinlaan 2, 1050 Brussels, Belgium.
Tl. +32 2 629 2129, Fax +32 2 629 2060, Email: averbeke@vub.ac.be

Isabel Wilmink
TNO–Inro, Schoemakerstraat 97, P.O. Box 6041, 2600 JA Delft, The Netherlands.
Tl. +31 15 269 6878, Fax +31 15 269 7782, Email: isw@inro.tno.nl

Preface

A growing interest of social scientists in general, and transportation experts in particular, in the European scene has been evident in the last ten years. This growing interest is due to actual scientific and applied developments in Europe, to the importance of the EC policies and research agenda, and to the trend of studying the difference between North American and European research. This interest also led to talks between the Network of European Communications and Transportation Activity Research (NECTAR) and American transport experts from the Institute of Transportation Studies (ITS) in California in order to formulate ways of cooperation. This book, apart from contributing to the general understanding of transportation systems, may thus serve as a window to present the four pertinent topics in the field of travel behaviour recently studied in Europe. Although it grew out of the fourth NECTAR Euroconference which took place in Israel (1998), it also includes additions written especially for it. In that respect it is wider than a post-conference collection of papers.

NECTAR is now a professional association which grew out of a European Science Foundation (ESF) funded network during the 1980s. The ESF can see in NECTAR a model in which the support of the networking has paid off. Leaving the nest of the ESF, NECTAR members have decided to continue both the meetings and the research collaboration. The current book is an outcome of this collaboration.

The book comprises four parts all based on non-traditional approaches. The first part deals with mobility and travel patterns. It analyses the relationship between housing and mobility, between infrastructure and mobility, and between parking and mobility. The second part examines the diverse range of relationships between telecommunication and travel, especially travel to work, and is complemented by an empirical study on long-distance telephone calls. The major issue of traffic congestion is presented and discussed in the third part. Special attention is given to the non-traditional concept that one should not get rid of congestion, but rather keep some. Some congestion is good for the system. Part III also presents several aspects of congestion on a European-wide scale as well as estimating the potential of traffic-calming measures, and assessing the consequences of urban sprawl from the point of view of environmental and social sustainability. Finally, the fourth part presents key issues in modelling behavioural responses. It presents innovations in the field of data collection as well as the notions of behavioural thresholds. It further presents a conceptual model of households' activity-scheduling process, based partially on empirical evidence gathered through a Computerized Household Activity Scheduling Elicitor (CHASE) survey. The book ends with questions concerning the possibility of modelling personal reactions to employment and residential relocations.

The editors wish to thank Mr Gregory Newman for his language and general editing assistance, and to acknowledge the graphical assistance offered by Mr Peter Lupan of the Department of Geography and Environmental Development at Ben-Gurion University. Special thanks are also due to Jennifer Chilton, Francine O'Sullivan and Julie Leppard of Edward Elgar Publishing for a fruitful cooperation throughout the publication procedures.

Eliahu Stern
Ilan Salomon
Piet H.L. Bovy

PART I

Mobility and Travel Patterns

1. The compact city: Conflict of interest between housing and mobility aims in the Netherlands

Kees Maat

1. INTRODUCTION

Urban land use has been dispersing and thinning out over the past few decades. For many, this implied that they could realize their dream of suburban living. While the countryside, including the 'Green Heart' of the Randstad metropolis, was being urbanized, the economic base of the urban facilities and services was being eroded. Moreover, the emerging spatial structure promoted the use of the private car and implicitly discouraged people from using bicycles and public transport. Consequently, the need for private cars kept on growing.

Already during the 1970s, opinions differed on whether the trend could be reversed. The reversal would mean promoting a more compact pattern of urbanization, thereby halting the trend toward a greater dependency on mobility, especially on the private car. That debate is still going on at the international scale. The Netherlands was one of the first countries to raise the topic. In fact, the Netherlands made the promotion of compact cities the cornerstone of its spatial planning policy. However, the government is now distancing itself from this principle.

As it turns out, the goal of curbing mobility by planning compact cities proves to be much harder to achieve than anticipated. In spite of all interventions to this end, the mobility rate is hardly diminishing. Nor are there signs of a shift from the use of cars to the use of bicycles and public transport. The difference between intent and reality, between policy and actual behavior, is generally described as the 'policy–behavior gap' (see Salomon and Mokhtarian, 1998).

In order to explain the emergence of this gap, it is useful to distinguish between two kinds of policy outcomes: instruments and effects (Maat and Louw, 1999). The *instrument* is the means to realize policy; in this case, the instrument is the model of compact urbanization. The *policy* effect refers to the kind of behavior that is to be encouraged. That is, the policy promotes the reduction of the total number of kilometers traveled and the shift from the use of cars toward the use of bicycles and public transport. On the one hand, the fact that the effects have not materialized must be blamed on the limitations of the theoretical notions on the effect of compact urbanization on mobility. This aspect is dealt with in section 5 below. On the other hand, the effects are disappointing because the policy to promote the compact city (the instrument) remains inadequate. It has been shown that the spatial and mobility objectives of the policy are largely incompatible with another goal, namely the promotion of private-sector housing. It is this conflict of interests that forms the core of the argument presented in this chapter.

2. HISTORICAL DEVELOPMENT

The process of suburbanization emerged when the fortifications of the cities lost their significance during the last century and transportation became easier. The spread of affluence after World War II brought car ownership within reach for growing numbers of people. With a car, they could live further away from their place of work, in locations that had hardly been accessible before. An enormous wave of suburbanization has engulfed the landscape since the 1960s, as more and more people could afford to leave the crowded cities and settle in rural intermediate zones. A house with a garden in the green setting of a village was more appealing to many people than an apartment in a newly constructed urban extension area. Moreover, the small-scale housing developments adjacent to the villages could meet the rising housing demand faster and for less money than new construction in the cities (De Boer, 1996). The effect was inevitable: the ongoing dispersal of population boosted the demand for automobile travel. But suburbanization also posed a threat to the population profile and the service facilities of the cities. Moreover, the expansion of the villages and the lines of new infrastructure broke the ground for the fragmentation of the countryside.

In a sense, the compact city principle can be seen as a backlash. This reaction is based on the assumption that the relationship between urbanization and mobility is reversible. Its motivation can be expressed as follows. If dwellings, jobs and services are located closer to each other, the distances that need to be covered will be shorter. In the aggregate, this should result in a smaller number of kilometers that need to be traveled. At the same time, greater proximity would boost the preference for the use of bicycles and public transportation.

A major incentive to hammer out a policy based on the concept of the compact city was the need to expand the housing stock. For years, a debate has raged in the United Kingdom between the proponents of compact cities and those who are counted as the supporters of 'new towns' (Breheny, 1996). This debate was fueled by the recognized need for 4.4 million new dwellings, a need that was expected to materialize between 1991 and 2016 (Williams and Banister, 1998). In the United States, the 'new urbanism movement' has set out to combat the continuing urban dispersal. This group of planners subscribes to a neo-traditional concept of development (see Ewing, 1997), a reference to the model of neighborhood development pre-dating World War II. Such residential areas are developed in higher densities and boast a mixture of functions, including shops within walking distance. Obviously, these neighborhoods contrast sharply with those built on the standard model for American postwar residential areas.

In the Netherlands, the concept of 'bundled deconcentration' became the model of choice for the urbanization pattern of the 1960s and 1970s. Dispersal was permitted, but attempts were made to control its effects. This mainly entailed efforts to concentrate the new construction in a limited number of places, namely in the 'growth centers' (Faludi and Van der Valk, 1992). The model grew out of the conviction that the cities themselves would be unable to satisfy the ever-growing demand for housing. During the 1970s, the implementation of the growth-center policy started to bear fruit. Concurrently, there was growing concern that the cities would keep thinning out while the countryside would fill up. Then the policy pendulum started to swing back towards meeting the needs of the established cities. The prevailing perception that the cities lacked space was overturned. Planners came to realize that there was enough room to build in the cities if densities were to be increased. During the 1980s, the National Physical Planning Agency commissioned a number of studies into the possibility of making the compact city concept a reality. With the publication of the Fourth Memorandum on Spatial Planning (Ministerie van VROM, 1988), the notion of the compact city was accepted as the foundation of spatial planning policy. At about the same time, it had become clear that a major thrust to increase the housing stock was needed once again. Consequently, the updated version of the memorandum, the Fourth Memorandum Extra (Ministerie van VROM, 1991), merged the new construction program with the compact city policy.

3. URBANIZATION AND ITS PRECONDITIONS

The housing need in the early 1990s amounted to approximately one million new dwellings, according to government estimates. The target for the construction program running from 1995 to 2005 was set at 634 800 units (Ministerie van VROM, 1996). Several hundred thousand more were to be built during the subsequent period.

Because of a variety of preconditions imposed on this new construction program by the state, its realization poses a complex challenge. The demand from the side of spatial planning was that the pattern of urbanization had to be compact, in part because of the need to curb the growth of mobility. For the same reason, the state has called for the construction of a high-quality public transport system (HOV) to serve the new residential areas. But at the same time, the state has formulated stringent accounting rules. These constraints are supposed to guarantee that the cost of providing public transport is covered by revenues from this service. With respect to housing, the condition was made that the new construction had to be financed predominantly by the private sector.

The Implementation of the Numerical Goals

The main purpose of housing policy is to ensure that the numerical targets are met. The policy objectives pertain to the stock: to make sure that planned dwellings are actually built; and to reduce the housing shortage. The implementation of that policy is the concern of local authorities and of the parties in the market (developers, private investors and housing associations). For the urban regions of seven large cities, the task of coordinating the implementation of the policy has been assigned to the so-called Framework Law Areas ('Kaderwetgebieden'). These are associations for intermunicipal cooperation within designated regions. For the remaining areas, the coordination is the responsibility of the provinces.

To guarantee the completion – and thereby the attainment of the numerical targets – the state has made 'hard' contractual implementation obligations for the period 1995–2005. The contracting parties are the state, the Framework Law Areas and the provinces. These entities have committed themselves to carry out the construction tasks. In so doing, they have accepted the responsibility of attaining the housing differentiation in accordance with the set conditions. To ease the development of the new residential areas, the state has offered to pay subsidies in exchange for strict implementation commitments. The state contribution amounts to approximately 9.5 billion guilders. The Ministry of Housing, Spatial Planning and the Environment (VROM) has allocated 2.5 billion guilders for the exploitation of the residential construction areas. The Ministry of Transportation and Public Works has dedicated over 7 billion guilders to the task of creating the public transport infrastructure that will serve these locations. Because of the contracts, the implementation has been safeguarded for a period of ten years. This guarantee precludes the need to hold negotiations on separate state support for each location.

The Ministry of VROM determined an exact set of conditions in its Decision on Location-linked Subsidies (BLS) for the state contribution that was meant to cover part of the cost of land. To ensure the timely completion of the dwellings, its contribution is paid in the form of a lump sum. The funds are given with no strings attached, apart from the condition that the residential construction should proceed according to schedule. The attainment of these output agreements is tested at two points in time: 1 January 2000 and 1 January 2005. As we shall see below, the Ministry of Transportation and Public Works has placed much more stringent conditions on the subsidies it is willing to offer.

Spatial and Mobility Conditions

To be able to meet the spatial and mobility conditions of this task, the Fourth Memorandum Extra contained detailed – and much more specific than before – designations of the locations of residential construction. These are called VINEX locations, VINEX being an acronym for VIerde Nota EXtra, the Fourth Memorandum Extra. About a quarter of the new dwellings will be built on small sites to increase the density of the existing urban areas; the remaining dwellings will be constructed in city extension schemes, outside the current built-up areas (Ministerie van VROM, 1996).

The two most important aspects of the VINEX policy are bundling and density. The bundling of the new residential locations with existing urban areas implies that the remaining space in the existing built-up area (BSG) is used first, locations at the edge of the city are the second priority, and construction in new extension areas some distance away from the cities comes in third place. This sequence of priorities expresses the basic assumption that a reduction of distance between the sites of activities will result in a reduction of mobility and that this, in turn, will make the bicycle more attractive as an alternative mode of transportation. Moreover, it is easier to connect close-in locations to existing public transport networks.

A higher residential density is not only necessary to promote a economical use of space and an increased demand for services and facilities. High density is also – and especially – important in promoting the use of public transport and the bicycle. Public transport needs a sufficient number of potential customers within easy reach of its stops (level of demand). The use of the bicycle is stimulated when the distances that a person needs to cover are short. The Ministry of VROM has therefore set the following minimum density norms: 30 to 35 dwellings per hectare in the Randstad, the urban west of the country; and 30 dwellings per hectare elsewhere.

Public Transport

Public transport is a key element of urban development policy. The VINEX policy envisions a 'leap in quality' in public transport, frequently using the adjective 'high-quality'. Initially, the Ministry of VROM considered urban fixed rail systems as the most suitable form for the Randstad region and separate rights of way for buses elsewhere. Afterwards, high-quality public transport (HOV) was defined as 'fast, frequent, reliable, and comfortable'. The particular mode to be chosen has become of secondary importance. For example, in the country's most extensive VINEX location, Leidsche Rijn near the city of Utrecht, the HOV will consist of a system of separate rights of way for buses.

The biggest investor in public transport in the VINEX locations is the Ministry of Transport and Public Works, which has committed itself to make substantial contributions. Moreover, for each of the public transport projects, the Ministry has agreed to cover potential shortfalls in the operating costs. These shortfalls are currently in the range of 60 to 70 percent and will therefore result in a major claim on the state budget. The growth of public transport will only increase these shortfalls in absolute terms. Therefore, the Ministry will seize the opportunity to test the cost-recovery potential of the VINEX locations before commitments are made for the financing of the projects. The level of cost recovery has to be shown to amount to at least 50 percent. This norm has led to an extensive involvement in the details of the future exploitation. By applying strict tests, an attempt is being made to force the conditions as much as possible toward an optimal recovery of the costs. Commitments for public transport investments of over 25 million guilders – which covers virtually all the projects in VINEX locations – will only be made if the plan has gone through all the steps of the Long-term Program Infrastructure and Transport (MIT). The MIT procedure is applied by the regional inspectorates of the Directorate-General for Public Works and Water Management (Rijkswaterstaat). It starts off with an informal orientation stage. After that, the projects proceed through an exploration stage, which entails drafting a traffic and transportation policy and conducting an engineering analysis of its objectives and necessity (this stage may last up to five years). Next comes a plan analysis stage in which alternatives for the problem solution and for the routing are evaluated (on average seven years). Only when these stages have been successfully completed – and only if the budgetary conditions can be met – is the implementation stage reached. At that point, the overall approval is given and the work can begin.

In addition to this, the state attempts to establish a businesslike operation by working through (privately owned) transport companies. These organizations reduce their operating shortfalls by applying market principles. The operating subsidies that are thereby freed up can then be transferred to investment.

Conditions for Residential Construction: The Swing Toward the Market

A state-imposed condition is that most of the residential construction is to be provided by the private sector, which will put the dwellings in the moderate or expensive range. At locations within the cities, at least 50 percent of the dwellings must be unsubsidized; in the extension areas, at least 70 percent have to be unsubsidized. Depending on local conditions, these minimum shares may be increased. In 1995 and 1996, respectively, 64 and 71 percent of the new construction was realized by the private sector; within the existing built-up areas of the cities, 64 percent was privately built (Ministerie van VROM, 1998).

The dependence on the market is based on the expectation that the private sector can achieve three intended effects. The first effect that the state aims to achieve is filtering within the existing housing stock. The main shortages are currently found in the most expensive housing segment. According to the official interpretation, the number of affordable dwellings is sufficient to house the low-income groups adequately. To some extent, the low-cost dwellings are occupied by households with a relatively high income (this situation is commonly described as the inexpensive misallocation). By promoting the construction of more expensive dwellings, the state hopes to promote upward filtering in the stock. Expensive dwellings tend to generate long vacancy chains. Consequently, they promote filtering; that is, more people tend to vacate an existing dwelling. To make this scenario succeed, the new dwellings will have to appeal to the potential buyers. Otherwise, those households will just remain where they are.

Second, dwellings in the market sector generate a larger income flow. The original intention was that the manipulation of the land prices could generate cross-subsidies to help finance the construction of social-rented dwellings. One problem is that the VINEX locations tend to be relatively expensive. They were left over from an earlier wave of urban development when the less expensive sites were developed. In the urban locations, the land prices and the cost to make them accessible are generally higher than elsewhere. In many places, extra money is needed to clean up existing soil pollution. In 1993, the price of land averaged 46 000 guilders per lot in a location at some distance from the city. The average price was 77 000 guilders at the edge of the city (the average VINEX location). Lots went for 64 000 guilders on average within the built-up area of the city. The lots in the average growth center cost 64 000 guilders (in 1993 currency). That amount included the contribution for the required extra infrastructure. In sum, lots in extension areas are roughly 20 percent more expensive than lots in the former growth centers. With respect to the occupants, it should be kept in mind that the purchaser of a dwelling in a growth center had to contribute 23 000 guilders toward the cost of the land. In contrast, someone who buys a dwelling in a VINEX location is charged on average 41 000 guilders (RPD, 1994). Finally, it should be realized that in many places the land prices have been forced up because of the activities of developers who bought into these project areas. They were able to buy the land cheap because the places that would be designated as VINEX locations had been publicized at an early stage. In fact, they were identified in the VINEX memorandum itself.

The third effect that the state is pursuing is an increase in quality. The VINEX policy aims to provide dwellings and environmental settings with future value. That is, the VINEX locations should remain in high demand for a long time. The private sector is assumed to be able to provide such a high-quality supply. The result should be an attractive housing stock.

4. THE DILEMMA: TO MEET RESIDENTS' PREFERENCES OR TO ATTAIN STATED MOBILITY GOALS?

A complex field of conflicting interests has emerged. There is tension between attaining the goal of market-oriented housing production and the goal of reducing mobility. To guarantee a sufficient demand for the dwellings, there is pressure to build in a less urban setting, one that is conducive to automobile travel. On the other hand, the goal of reducing overall mobility requires a more compact urbanization pattern.

The successful completion of the construction assignment in VINEX locations is ultimately not determined by public action. Rather, it depends on the willingness of the public to purchase the dwellings (Kruythoff, 1996). That willingness can only arise if the potential for residential satisfaction in the new locations is greater than that generated in the current or alternative locations. This is certainly the case if people are supposed to filter up. The picture of what the residential consumer wants has become clear from various investigations (Wassenberg et al., 1994; Rietdijk and Boelhouwer, 1997). Most homebuyers would like to own a single-family dwelling, preferably not built as an attached dwelling in a row; a garden is the most preferred 'amenity.' A suburban or even a rural residential setting is preferred above an urban neighborhood near facilities and services. This is also the preference of two-income households without children. Moreover, the inhabitants of VINEX locations prove to have a strong attachment to their cars. Indeed, they are reluctant to trade off access to parking space against access to public transport. The rate of car ownership is above average among the target group: 90 percent of the households own a car, while 23 percent own at least two vehicles.

The providers of housing in the market fear that the relationship between price and quality will become less advantageous. If that were actually the case, there would be some danger that not all dwellings could be sold when the market becomes more relaxed. The construction cost of housing in VINEX locations is relatively high; the price differential with housing elsewhere is hardly offset by subsidies any more. The consumer will want to have a superior dwelling in a more desirable setting in return for this higher price. This in itself pushes up the prices even further. It should therefore come as no surprise that at this high price level, mostly relatively expensive homeowner dwellings are being built. Partly because of the currently low interest rates and partly because the market remains tight, there is a sufficiently large number of buyers. But the market conditions will certainly change in the near future. Specifically, the market will gradually become much more relaxed – which, by the way, is in accordance with the goals of the VINEX policy.

Although the prices at the VINEX locations remain relatively high, the quality tends to be lower than in other locations for new construction in several respects. An investigation conducted by the Dutch Association of Homeowners (Vereniging Eigen Huis, 1997) has revealed that the residential lots average 170 square meters. That is one-third less than for the newly built housing in other locations. Moreover, many new dwellings have been built in rows and have neither a carport nor a garage. Elsewhere, there are many more new free-standing or linked dwellings, and they include the coveted garage!

This relatively unfavorable position may seriously affect the competitive position of the VINEX locations. This is all the more serious because these locations still retain a huge planned capacity in reserve: the current structure and land-use plans allow for an additional 169 000 dwellings to be constructed at the very least (Keers and Van Heuven, 1997). The smaller municipalities especially have a vested interest in boosting their population size to increase the customer base for their services and facilities. The sale of building lots also generates substantial income for the local authorities. It should be kept in mind that the small municipalities, with their small developments, suffer less from long-drawn-out administrative procedures and from soil pollution. Moreover, they do not need to provide expensive infrastructure for access. Under these favorable conditions, they can rapidly proceed with their residential construction plans. The resulting 'green', suburban settings prove to be captivating to many households that have a potential interest in trading up (Priemus, 1998).

Meanwhile, it can be shown that the attainment of a market orientation for the housing in the VINEX locations provides a sharp contrast to the mobility goals. The selected locations were preferred because they were close to existing urban areas. This also implies that many of them are favorably situated close to the major traffic arteries. For instance, Leidsche Rijn is near the intersection of two main national expressways, providing connections to the four biggest cities of the country (Teisman, 1997); Ypenburg near The Hague has an optimal location with respect to the major expressways, namely the A4, the A12 and the A13. Conversely, the VINEX locations are way outside the reach of existing railway stations. The distance to the cities is too large to induce people to commute by bicycle. The effect of distance is aggravated by the barriers of the existing intervening expressways.

The high-quality public transport – intended to offer strong competition to the automobile and to form an asset that would boost the quality of the residential environment – is taking a long time to materialize. The condition set by the Ministry of Transport and Public Works – that the fares must cover at least 50 percent of the cost of operation – proves to be too much of a hindrance for many of the projects. In a large number of the VINEX locations, the layout and the density prevent this demand from being met. Densities of 30 to 35 dwellings per hectare are insufficient to support high-quality public transport. Many of the developments don't even come up to these minimal densities. Only in a limited number of intra-city locations have such densities actually been achieved. These central locations, however, have been targeted to meet the preferences of a minority of the urban-oriented clientele (20 to 25 percent). Apart from the insufficient density, the layout often proves inadequate to allow a sufficient cost-recovery rate. This aspect caused the government to fine-tune its policy. The update of the VINEX (Ministerie van VROM, 1997) therefore states that it is not only the distances between the dwellings, job locations, and facilities and services that are important. The layout of the locations should also stimulate the use of bicycles and public transport. An example of a development where this has not happened is the VINEX location Meerhoven near the city of Eindhoven. There, the houses have been sited around a small lake – because living on the water's edge is considered to be attractive. Lying close to a beltway, the dwellings are well suited for automobile users. However, the HOV route skirts the area. It is only within easy reach for the people living on one side of the lake (Priemus and Maat, 1998).

As a result of the lack of compactness in the design, only a few HOV plans have completed the MIT process and have had their financing (fully) approved. Meanwhile, residential construction continues on schedule. Thus, many transport projects will either not be completed or will become operational too late. This could turn out to be a disaster for the policy aim to curb automobile use. Once people have chosen the car, they do not readily go back to public transport. When individuals make an important decision – such as the initial selection of where they will live or work, or such as their later decision to change their place of residence or employment – the context for their mobility becomes fixed for a long period of time. Such deliberate long-term choices provoke a reconsideration of short-term mobility behavior, such as the choice of the mode of transport. Afterwards, when such choices are no longer explicit, the behavior becomes a routine (Salomon and Ben-Akiva, 1983).

The dilemma assumes the following form. The state wants to put an end to the housing shortage. But at the same time, it sets conditions for mobility, the use of space and the claim on the public budget. The spatial and mobility goals suggest a compact urbanization model at locations within or close to existing cities, where high-quality public transport is available. The aim to control public expenditure steers the government toward the market, both for housing and public transport. With respect to housing, this entices the providers to cater especially to the residential preferences of potential purchasers.

The goal of developing truly compact neighborhoods with an optimal design for bicycles and public transport is, however, incompatible with the residential preferences of most of the people who buy expensive homes. Apart from a limited number of centrally located sites, the competitive position of dwellings is undermined when the areas are closer in, more compact, and less automobile-friendly because the dwellings are thereby further removed from the housing aspirations of potential buyers. Their position is weakened even more when high costs cause the price–quality ratio to deteriorate. Finally, the weakness of the competitive position is brought into sharp relief when compact neighborhoods are compared with the housing supply in suburban locations.

Many developers are therefore inclined to build neighborhoods that look as much as possible like suburban areas. They want to build at the lowest possible densities and provide optimal access for the automobile. The municipalities in question are eager to cooperate. Buyers with an above-average income are very welcome as new members of the community. Moreover, the municipalities have obliged themselves to meet the targets stipulated in the VINEX contracts. If the numbers are not reached, or not in time, the state may stop the payment of the BLS subsidies or even claim paid amounts back (after 2005).

On the other hand, areas with a suburban flavor and good access for automobiles generate an insufficient level of demand for efficient public transport. Insufficient demand, in turn, slows the provision of HOV or can result in the cancellation of plans to put in the system. The norms of the MIT process carried out by the Ministry of Transport and Public Works are so strict that few projects have obtained financial support as yet. This leads to major delays in the realization of the entire transportation concept.

5. THE EFFECTS OF INCREASE IN DENSITY

Are there ways to resolve the dilemma? Is it possible to have the best of both worlds? What are the options to reach the homebuilding targets as well as the mobility targets? To be able to answer these questions, the effectiveness of the compact-city policy for the attainment of the mobility goal should be evaluated.

Although the idea behind the compact-city model sounds plausible, the real world proves to be much more resistant to change than it is often assumed to be. Many international studies (Crane, 1996; Kitamura et al., 1997; Banister et al., 1997; Cervero and Kockelman, 1997) have found some evidence of a relationship between urbanization and mobility behavior. But the level of elasticity of the urbanization factors proves to be low. In other words, a strong increase of compactness results in only a limited reduction of the number of kilometers traveled by car.

The Dutch situation, however, is hardly comparable to the American one. Unfortunately, studies on the Netherlands are scarce. Den Hollander et al. (1996) have evaluated the impact of the proximity principle on the commuting rate by comparing the behavior before and after a move to a new residential location. The new residential locations resembled the VINEX locations in the Randstad, all being accessible by public transport. The research showed that with a use rate of 60 percent, the car was and remained the preferred mode of transport. Furthermore, the use of the car showed hardly any tendency to decrease. The average journey to work has become slightly shorter. However, any reductions were entirely accounted for by a small number of households who had moved over a very long distance and had reduced their commute substantially. The other households proved to have increased the distances they covered. Konings et al. (1996) investigated approximately 25 new residential locations in the province of Noord-Brabant. They compared the travel behavior before and after the move. In doing so, they distinguished three classes of locations (intra-urban, urban extensions and rural) and three density classes. That exercise revealed that after the move there was no reduction in the length of the commutes at all. (Just as in the Randstad, the only reduction was brought about by a limited number of long-distance movers, while the others traveled further.) Their study also showed that the use of the car remained at a very high level. Even in combination with a residential relocation, travel behavior proves hard to change. None the less, a weak yet significant relationship with residential density did show up: the use of the car declines slightly where the density of the development increases But the use of public transport remains disappointingly low. The bicycle, however, plays a remarkably large role, especially in the rural areas.

Neither of the two investigations offers convincing proof that the compact-city policy will be successful (Maat, 2001). How can this be explained?

There are various developments that complicate the application of the compact-city model.

First, the characteristics of the households exert a major influence. Affluent households in new residential locations are strongly geared to automobile use and account for much of the overall mobility. Thus, it seems likely that individuals or households with particular lifestyles generate more trips by car than households with different lifestyles – independent of their residential location. This corresponds with the findings of Kitamura et al. (1997). They indicate that effects of a compact-city policy on mobility patterns are visible, but that personal attitudes are more strongly associated with mobility behavior than characteristics of the urbanization pattern.

Another reason for the lower-than-expected effects of the compact-city policy is that the artifact (the physical environment) is confused with the activity. The patterning of the city on the proximity principle is supposed to guarantee that many activities will take place within the context of the urban region. Yet several factors prevent the adaptation of the activity patterns; there is a clear trend toward what may be called the distancing of activities. The following examples serve to illustrate this development. The labor market reveals a tendency to increasing specialization, which results in the enlargement of the search space for jobs. Moreover, labor mobility has been increasing, but this does not entice people to look for a different dwelling near their new job. In a qualitative sense, the housing market cannot respond adequately to the changes in demand. An attractive dwelling or a desirable suburban residential setting is often used as a justification for a longer commute. Two-earner households generate more journeys to work than households with a single wage-earner. Moreover, the trips made by the members of two-earner households tend to be longer. That is because, when selecting a residential location, they have to take account of two work locations. Consequently, they select either a place near one of the jobs or a place lying in between the two. We therefore share Banister's (1992) conclusion that the 'notion of a compact city with a greater diversity and people living close to their workplaces simplifies the complexity of lifestyles.'

The compact-city model is based on the assumption that mobility is essentially dependent on other causes. The demand for trips is caused by individuals' need to participate in other activities related to home or work and the use of facilities. In addition, such activities generate trips because they tend to be geographically separated. Hence, by changing the spatial pattern of activities – as done in the compact city – mobility would be reduced. But this outcome hinges upon the assumption that individuals will use the opportunity to reduce their travel if they are given that option. Salomon and Mokhtarian (1998) argue that there are sufficient indications of flaws in that assumption. One of those indications is that people will almost always convert a gain in travel time (for instance through the use of a faster transport mode) into an increase of the distance covered (Schafer and Victor, 1997). Likewise, it is not unlikely that people who live in compact cities will use the time saved by not having to commute to make other trips.

It is conceivable that compact cities will even generate extra mobility. There are some indications that people demonstrate substitution behavior. That is, they will seek compensation for their lack of space elsewhere within the city. It is also probable that compact cities – with their large population and short internal distances – could maintain efficient interurban connections. If so, this would boost the number of long-distance trips.

6. CONCLUSIONS AND PERSPECTIVE

In the VINEX policy memorandum, the program for urbanization is embedded in a vision for the realization of compact cities. The choice for a model of compact urbanization is largely based on past experience. Nevertheless, it has turned out to be extremely difficult to implement such a model. Compact urbanization appears to conflict with the preferences of the potential inhabitants. An increase in residential density reduces the city's attractiveness and encourages those who can afford it to move away.

One of the results of the failure of compact urbanization to materialize is that the demand for (high-quality) public transport remains inadequate. This was the reason why until recently the Ministry of Transport and Public Works was reluctant to make any financial contributions to such projects. However, although the density and the urban design of new residential locations have not been improved, the Cabinet has recently decided to make extra investments in the public transport projects in VINEX locations (Ministerie van Verkeer en Waterstaat, 1998). These extra investments suggest that the principle of cost recovery from fares has become less important than before. Moreover, the delay has become so long that by the time public transport is actually provided – probably in a few years – most of the inhabitants will have developed a pattern of mobility focused on their car.

Because it proves to be so difficult to realize the compact city, it would only make sense to continue to press ahead with the policy if it can be shown to yield major benefits in terms of a reduction of (auto)mobility. This, however, is not the case. The assumed mobility effects of compact cities have been shown to be minor. It is therefore useful to keep searching for other models of spatial urbanization that will reduce travel by car, but, at the same time, allow for the creation of residential environments where people want to live.

This poses planners a difficult challenge. Most people want to live in a green environment. Yet present and past experience has convincingly shown that suburbanization increases the demand for mobility. The Ministry of VROM (1999) seems to be moving toward a preference for corridors and network cities. Both forms of development can make optimal use of existing infrastructure. But it remains to be seen whether that will also engender enormous traffic flows. It is hoped that public transport will be able to provide the means to accommodate the extra flows of people.

It might be prudent to utilize the potential of the bicycle by taking specific measures in the urban environment. The town of Houten provides a good example of how this approach might turn out. The town provides suburban-type residential environments. But in spite of the high rate of car ownership, the number of trips by car is relatively low, while people use the bicycle substantially more than elsewhere. The layout of Houten is especially cyclist-friendly: direct bicycle routes connect the residential neighborhoods to the town center and the railway station. Facilities for recreation and social services are as much as possible linked directly into the backbone of the network of bicycle routes. The centrally located railway station is within 1600 meters of the dwellings that are furthest from it. The station itself contains an abundance of provisions for cyclists. Cars are allowed to penetrate into the core of the town but are otherwise relegated to an orbital road. In short, an urban design that is above all friendly to bicycle use and public transport might be able to resolve the dilemma.

REFERENCES

Banister, D. (1992), 'Energy Use, Transport and Settlement Patterns', in M.J. Breheny (ed.), *Sustainable Development and Urban Form*, London: Pion.

Banister, D., S. Watson and C. Wood (1997), 'Sustainable Cities: Transport, Energy, and Urban Form', *Environment and Planning B: Planning and Design*, **24**, 125–43.

Breheny, M. (1996), 'Centrists, Decentrists and Compromisers: Views on the Future of Urban Form', *The Compact City: A Sustainable Urban Form?* London: Spon, pp. 13–55.

Boer, N. de (1996), *De Randstad bestaat niet. De onmacht tot grootstedelijk beleid*, Rotterdam: NAi Uitgevers.

Cervero, R. and K. Kockelman (1997), 'Travel Demand and the 3 Ds: Density, Diversity and Design', *Transportation Research D*, **2**, 199–219.

Crane, R. (1996), 'Cars and Drivers in the New Suburbs: Linking Access to Travel in Neotraditional Planning', *Journal of the American Planning Association*, **62**(1), 51–65.

Ewing, R. (1997), 'Is Los Angeles Style Sprawl Desirable?', *Journal of the American Planning Association*, **63**(1), 107–26.

Faludi, A. and A. van der Valk (1992), 'Growth Regions and the Future of Dutch Planning Doctrine', in M.J. Breheny (ed.), *Sustainable Development and Urban Form*, London: Pion, 122–37.

Hollander, B. den, H. Kruythoff and R. Teule (1996), *Woningbouw op Vinex-locaties: effect op het woonwerk-verkeer in de Randstad*, Delft: Delft University Press.
Keers, G. and J. van Heuven (1997), *Concurrentie in woningproductie tussen Vinex-bouwgemeenten en buitengebied*, Amsterdam: Rigo.
Kitamura, R., P. Mokhtarian and L. Laidet (1997), 'A Micro-analysis of Land-use and Travel in Five Neighborhoods in the San Francisco Bay Area', *Transportation*, **24**, 125–58.
Konings, J.W., H.M. Kruythoff and C. Maat (1996), *Woningdichtheid en mobiliteit. Woon-werkverkeer op nieuwbouwlocaties in de provincie Noord-Brabant*, Delft: Delft University Press.
Kruythoff, H.M. (1996), 'Woonconsument en inrichting van Vinex-locaties', in F.M. Dieleman and H. Priemus (eds), '*De inrichting van stedelijke regio's*,' Assen: Van Gorcum.
Maat, K. (2001), 'Effects of the Dutch Compact City Policy on Travel Behaviour', in E.T. Verhoef and E. Feitelson, *Transport and Environment: In Search of Sustainable Solutions*, Cheltenham: Edward Elgar.
Maat, C. and E. Louw (1999), 'Mind the Gap: Pitfalls of Travel Reduction Measures', *Built Environment*, **25**(2), 151–61.
Ministerie van Verkeer en Waterstaat (1998), *Meerjarenprogramma Infrastructuur en Transport 1999–2003*, Den Haag: Sdu Uitgevers.
Ministerie van VROM (1988), *Vierde Nota over de Ruimtelijke Ordening* (Fourth Report on Spatial Planning), Ministry of Housing, Spatial Planning and the Environment, Den Haag.
Ministerie van VROM (1991), *Vierde Nota over de Ruimtelijke Ordening Extra*, Den Haag: Sdu Uitgevers.
Ministerie van VROM (1996), 'Verstedelijking in Nederland 1995–2005', *De Vinex-afspraken in beeld*, Den Haag: Sdu Uitgevers.
Ministerie van VROM (1997), *Actualisering Vierde Nota over de Ruimtelijke Ordening Extra*, Den Haag: Sdu Uitgevers.
Ministerie van VROM (1998), *Rapportage monitoring uitvoering verstedelijking Vinex 1998*, Den Haag: Sdu Vitgevers.
Ministerie van VROM (1999), 'De ruimte van Nederland', *Startnota ruimtelijke ordening 1999*, Den Haag: Sdu Uitgevers.
Priemus, H. (1998), 'Op zoek naar een nieuw ruimtelijk integratiekader', *Stedebouw en ruimtelijke ordening*, **1**, 16–21.
Priemus, H. and C. Maat (1998), 'Ruimtelijk en Mobiliteitsbeleid: interactie van rijksinstrumenten', *Stedelijke en Regionale Verkenningen 18*, Delft: Delft University Press.
Rietdijk, N. and P. Boelhouwer (1997), '*Huizenkopers in profiel, wie zijn zij en wat willen zij?*', Voorburg: Nederlandse Vereniging van Bouwondernemers.
Rijksplanologische Dienst (1994), *De prijs van een plek*, Den Haag: Ministerie van VROM.
Salomon, I. and M. Ben-Akiva (1983), 'The Use of Life-style Concept in Travel Demand Models', *Environment and Planning A*, **15**, 623–38.
Saloman, I. and P.L. Mokhtarian (1998), 'What happens when mobility-inclined market segments face accessibility-enhancing policies?', *Transportation Research D*, **3**, 129–40.
Schafer, A. and D. Victor (1997), 'The Past and Future of Global Mobility', *Scientific American*, October, 58–61.
Teisman, G.R. (1997), *Sturen via creatieve concurrentie, een innovatie-planologisch*

perspectief op ruimtelijke investeringsprojecten, Nijmegen: Katholieke Universiteit Nijmegen.

Vereniging Eigen Huis (1997), 'Aanbod op de grote nieuwbouwlocaties', *Woonconsument*, **12**, 7–8.

Wassenberg, F.A.G., H.M. Kruythoff, T.A.L. Leliveld and J.E.H. van der Heijde (1994), *Woonwensen en realisatie van Vinex-locaties in de Randstad*, Den Haag: Ministerie van VROM.

Williams, J. and D. Banister (1998), 'Future Sustainable Housing Strategies', paper presented at *Housing Futures: Renewal Sustainability and Innovation*, September 1998.

2. Transport networks and mobility: A comparison analysis of the Randstad, the Rhine–Ruhr area and the Antwerp–Brussels–Ghent region

Gysbertus R.M. Jansen, Hans Hilbers and Isabel Wilmink

1. BACKGROUND AND OBJECTIVES

The traffic volumes of the Dutch motorway system are very high. In particular, the motorways in the highly urbanized and densely populated regions in the western and central part of the country are trafficked heavily, resulting in a high level of congestion. The number of these tailbacks, as well as their duration, has increased steadily during recent years.

A key objective of Dutch economic policy is to provide and maintain high accessibility and a reliable infrastructure in order to enhance its competitiveness in a European context. In addition to accessibility as a locational factor for businesses in general, the importance of the transport and freight sectors within the economy may explain this strategy.

The increasing level of congestion is thought to result in a deteriorating level of accessibility, which in turn leads to a lower level of economic growth and employment. Since the central cities of the Randstad area in particular are suffering from high levels of unemployment and congestion, there was a strong desire to carry out a comparative analysis of the mobility patterns, transport supply and accessibility levels in competing and comparable European metropolitan areas. A number of interrelated projects have been commissioned by the Dutch Ministry of Transport and Public Works to TNO–Inro including performing these comparisons for personal travel (Hilbers et al., 1996; 1997; Hendriks et al., 1997).

Since the Randstad is a multimodal urban area, it was compared with similar areas within the same European region, which, in theory, are competing for the same business. The Rhine–Ruhr (Ruhr) area in Germany and the Antwerp–Brussels–Ghent (ABG) region in Belgium were selected as the comparative cases.

The general goal was to assess in a comparative manner the accessibility and the levels of service of the transport systems in these three areas. The specific objectives were:

- to compare empirical traffic volumes
- to compare the supply of transport and the transportation network structure
- to compare mobility levels
- to determine the key distinction factors such as urban structure, housing and labour markets
- to investigate investment programmes and volumes for strategic transport facilities
- to compare planning and decision-making procedures.

The last two objectives address the concern that Dutch levels of investment are low and that planning procedures are too slow. It should be noted that a direct comparison of congestion levels was not possible within the projects that have been carried out, mainly because of a lack of consistent data within these regions. The original studies had a clear multimodal focus dealing with both car traffic and public transport. This chapter will mainly address automobiles.

The organization of the chapter is as follows: section 2 deals with the data sources and the delimitation of the study areas. Section 3 summarizes the comparison of traffic volumes. Section 4 describes road transport supply. It is followed in section 5 by an analysis of the distribution of traffic volumes along the different classes of roads. Section 6 looks at mobility patterns as a background factor in the differences in traffic volumes. Section 7 examines the investment programmes and touches on decision-making procedures. Finally, section 8 summarizes the main findings. This chapter gives only a selective and brief summary of the results of a multi-year effort. We refer to the reports and other publications listed at the end of the chapter for more information.

2. DATA SOURCES AND GENERAL CHARACTERISTICS

Data

It was beyond the scope of the studies to collect new data. Data from official and other public sources have been used. Specific analyses have been made using the data files of the mobility surveys in the Netherlands (OVG) and Germany (KONTIV). The analyses of the road networks have been made using the classification methodology and maps published by Michelin.

Area Delimitation

Extensive study has been made of the criteria and methods to define the three areas in a consistent manner. We refer readers to Hilbers et al. (1996) for a detailed description of the approach followed. It is useful to note that in our study, the Ruhr area has been defined to extend beyond the core area between Duisburg and Dortmund to include the cities of Dusseldorf, Cologne and Bonn.

Some General Characteristics

Table 2.1 presents some general characteristics related to area, population, employment and transport networks. We will briefly highlight the differences between the Randstad and both the Ruhr area and the ABG region.

The Ruhr area is twice as large as the Randstad. However, the population density is higher in the Randstad. A marked difference between these two areas is that, unlike in the Randstad, a large number of the cities in the Ruhr area have grown into one contiguous urbanized area. As a result of this urban form, the motorway network varies between these two regions. In the Ruhr area, more parallel connections are available in the central part. On a per capita or per km^2 basis total length is similar. The rail network in the Ruhr area is more dense; however, since it provides fewer trains the service level is lower.

The ABG area is slightly larger than the Randstad, but both the population and the population density are smaller. A striking characteristic of the ABG area is its small number of medium-sized cities. It can be concluded that the Randstad has the smallest area and the highest population density of these study areas. The motorway network in the ABG area has a clear structure with ring roads around the three main cities and connections from these cities radiating outwards into the surrounding areas. The total length is about equal to that of the Randstad network. The railway network in the ABG region is very dense, but low service frequencies result in an equal number of train kilometres per capita with the Randstad, although the number of railway stations is much higher.

3. TRAFFIC LOADS OF THE ROAD NETWORK

Traffic volume is used as a proxy variable for congestion level since no readily available and consistent data on congestion were available. It is necessary to express volume as a ratio to capacity; Table 2.2 shows the average volume per lane for the road network in the three areas.

Table 2.1 General characteristics of the three areas Randstad, Ruhr and ABG

	Randstad	Ruhr area	ABG area	Randstad compared to Ruhr area (%)	Randstad compared to ABG (%)
Area (km^2)	5200	12 000	5500	–57	–5
Inhabitants (mn)	6	11.8	4.3	–49	+40
Jobs (mn)	2.3	5.0		–54	
Inhabitants per km^2	1150	980	780	+17	+48
Jobs per km^2	440	420		+6	
Motorway network (km)	640	1430	470	–55	+36
Railway network (km)	700	1820	1030	–61	–32
Railway stations	131	417	215	–69	–39
Train kilometres per day	123 000	195 000	88 000	–37	+40
Train departures per day	18 000	42 000	15 000	–57	+20
Average speed of trains (km/h)	63	59	60		
Average distance between stops (km)	6.8	4.6	5.8		
Network tramway/underground (km)	320	960	250	–66	+29
Vehicle km tramway/underground per day	94 000	168 000	62 000	–44	+52

Table 2.2 Average daily traffic volumes per lane by functional class (vehicles/24 hour day; weekdays only)

Daily traffic volumes per lane	Randstad	Ruhr area	ABG area
Motorways	16 800	13 200	13 600
Other main roads	5 900	5 700	6 200
Regional roads	4 800	3 800	3 800
All roads	10 000	8 100	8 000

This table reveals that the average motorway traffic volume per lane in the Randstad is substantially higher than in both other areas. This difference of about 20 per cent indicates a significantly higher level of congestion in the Randstad since at high load/capacity ratios the congestion level increases exponentially with volume. A similar difference can be seen with the regional roads. However, given the lower volume/capacity ratios on these roads, this difference will probably not lead to a much higher level of congestion.

Looking into greater geographical detail, our analysis shows that within the Randstad the motorway links near the four larger cities (Amsterdam, Rotterdam, The Hague and Utrecht) have the largest per-lane traffic loads. In addition, the crucial intercity connections (Amsterdam–Utrecht, Amsterdam–The Hague and Rotterdam–The Hague) are loaded extremely heavily along almost their complete length. In the motorway network, no parallel alternatives are available, which demonstrates the vulnerability of the system in this respect. The connections of the Randstad with its surrounding areas are less heavily loaded in most cases.

In the Ruhr area too, the most heavily loaded motorways can be found around and between the main cities. Compared to the Randstad, the average volume/capacity ratio is lower on these types of connections. For example, the average daily traffic volume on orbital motorways around the German cities is about 16 300 vehicles per lane and circa 20 100 on similar stretches in the Randstad. The intercity motorways in the Ruhr area are slightly less loaded: 15 600 vehicles versus 17 700. We will deal with the possible causes of these differences later in the chapter.

In the ABG area the highest congestion levels can be found on the ring roads around Brussels and Antwerp. The per-lane volumes on intercity connections are not high. A convenient way to indicate the congestion levels on motorways is the proportion of motorway length with a daily volume of 20 000 vehicles per lane. These sections comprise 23 per cent in the Randstad, 12 per cent in the Ruhr area and 5 per cent in the ABG area, respectively. These figures confirm the concern underlying this study that congestion problems in the Randstad are higher than in similar multimodal conurbations.

4. SUPPLY OF INFRASTRUCTURE

One of the factors determining traffic load is the available supply of roads. In this chapter we will analyse the amount and characteristics of the road networks. Mobility patterns, which are also affected by road supply, are dealt with later.

Two measures are used to quantify road supply: road length per area unit and lane length per inhabitant. The first measure expresses the network density in the region. As Table 2.3 shows, the density of the total road network is equal in the Randstad and the ABG area and higher in the Ruhr area. Differentiating by type of road, it can be seen in Table 2.3 that the density of the motorway network is similar in the Randstad and the Ruhr area and much lower in the ABG area.

Table 2.3 Network density and capacity of main road network

	Randstad	Ruhr area	ABG area
Network density (road length in km per 1000 km^2)			
Motorways	115	120	80
Other main roads	105	185	180
Regional roads	180	150	140
All roads	400	455	400
Capacity (lane length in km per 1 000 000 inhabitants)			
Motorways	480	520	570
Other main roads	320	460	690
Regional roads	380	370	450
All roads	1190	1360	1710

A striking feature of the Randstad is the very low density of non-motorway surface roads. The density is only about half of that in the other regions. It can be concluded that in the Randstad almost no connections of a functional level just below that of motorways are available.

If population density is taken into account as a first indicator of transport demand, we notice in Table 2.3 that capacity measured as lane length per capita reveals a different pattern. Total road supply is lowest in the Randstad: about 13 per cent lower than in the Ruhr area and about 40 per cent less than in the ABG area. The supply of motorways is lowest in the Randstad: about 8 per cent lower than in the Ruhr area and 15 per cent lower than in the ABG area. The relative absence of non-motorway main roads in the Randstad is revealed very clearly by this measure.

It may, therefore, be assumed that the higher traffic volumes and the higher levels of congestion in the Randstad can be explained, to a certain extent, by the lower level of road supply. The almost complete lack of a lower-level surface network in this area has significant consequences for the vulnerability of the total system to congestion and incidents. In most cases, only one motorway connection is available and no lower-level surface alternatives exist.

5. DISTRIBUTION OF CAR KILOMETRES BY FUNCTIONAL CLASS OF ROADS

One of the hypotheses was that given the lack of a substantial lower-level road network and the short distances between exits on motorways[1] in the Randstad, motorways in this area are being used more for shorter trips than in both other areas. This theory could also explain the high traffic volumes on those links.

To test this assumption two analyses have been made:

- the distribution of car mileage by type of road
- the share of the motorway segment by trip distance class.

Table 2.4 shows the distance travelled by car[2] on various types of road per capita. It is clear from the analysis of these areas that the per capita

Table 2.4 Distribution of daily distance travelled by car per capita on various types of road (km)

	Randstad		Ruhr area		ABG area	
	km	%	km	%	km	%
Motorways	8.1	68	6.9	63	7.7	56
Other main roads	1.9	16	2.6	24	4.3	31
Regional roads	1.8	15	1.4	13	1.7	12
All roads	11.9	100	10.9	100	13.8	100

distance travelled by car on motorways is largest in the Randstad. This finding is partly due to the higher total level of car use, but to a greater extent it is caused by the larger overall use of the motorways. The lack of a well-developed underlying road network indeed results in a larger use of motorways.

The second question raised above was whether in the Randstad shorter trips are made using motorways compared to both other regions. To answer this question two items have been established:

- the share of the motorway in the distance travelled by trip length
- the distribution of total car kilometrage by trip length.

Table 2.5 shows the outcomes of both analyses for the Randstad and the Ruhr area.

Table 2.5 Distribution by trip distance class of car kilometrage and share of motorways in distance travelled for each trip distance class

Trip distance class (km)	Share of distance class in total car kilometrage (%)		Share of motorway in kilometrage travelled in each distance class (%)		Motorway car kilometrage by distance class (%)	
	Randstad	Ruhr area	Randstad	Ruhr area	Randstad	Ruhr area
0–15	28	36	25	27	12	16
15–30	22	19	56	56	21	19
30–50	16	14	70	71	19	17
50–100	19	12	83	84	27	17
100+	14	20	92	92	22	31
Total	100	100	60	58	100	100

The figures show that the mode share of the motorway for each trip distance class is almost identical. As was expected, the longer the trips the larger the proportion of the trip travelled on motorways. The hypothesis that shorter trips in the Randstad are made using motorways to a larger extent proves to be false.

If the car kilometrage on motorways is analysed by trip distance class, it is seen in Table 2.7 that the share of the short trips is even smaller in the Randstad. This has to do with the shorter total distance travelled on these trips.

6. CHARACTERISTICS AND FACTORS OF MOBILITY PATTERNS

Mobility Pattern

Differences in traffic loads may be explained by differences in mobility in the three areas. Personal mobility characteristics of the inhabitants of the Randstad and the Ruhr area have been analysed using the 1989 travel survey data from OVG and KONTIV. The results of the personal travel survey have been added for comparison wherever possible.[3] The findings are summarized briefly below with respect to the distance travelled per capita and the share of peak period travel.

Table 2.6 shows that distance travelled per day per capita is highest in the Randstad and similar to the Flemish figure. The residents of the Ruhr area travel significantly less distance.[4, 5] The distribution by purpose shows that the travel is particularly less for work and leisure purposes.

Table 2.6 Total distance travelled per capita by purpose

	Randstad		Ruhr area		Flanders	
Trip purpose	km	%	km	%	km	%
Work	8.3	25	6.6	29	8.4	25
Business	3.2	10	2.4	11	3.4	10
Education	1.5	5	1.4	6	1.7	5
Shopping	3.5	11	3.2	14	4.2	12
Leisure and social	12.6	39	7.3	32	13.7	40
Other	3.4	10	1.6	7	2.9	9
All trips	32.5	100	22.5	100	34.4	100

Table 2.7 gives the distribution of the total distance travelled by mode of transport, showing slight differences in the modal split. The total share of the car mode is remarkably similar in all three areas, ranging between 72 and 78 per cent. It can be seen that the share of the bicycle mode in the Randstad is highest, as might be expected.

From analyses not shown here, it can be concluded that mobility is strongly related to socio-economic categories. Both in the Randstad and the Ruhr area it was found that full-time workers travel much more than students, housewives and the elderly. A marked difference is that inhabitants of the large cities have the lowest mobility level in the Randstad, whereas these groups

Table 2.7 Total distance travelled per capita by mode of transport

	Randstad		Ruhr area		Flanders	
Mode of transport	km	%	km	%	km	%
Car driver	16.4	50	13.0	58	17.8	52
Car passenger	7.1	22	3.8	17	9.0	26
Public transport	4.7	14	3.6	16	3.8	11
Bicycle	2.9	9	0.8	4	1.7	5
Walk	0.8	2	1.1	5	0.4	1
Other	0.6	2	0.2	1	1.6	5
All trips	32.5	100	22.5	100	34.4	100

have the highest level in the Ruhr area. It is hypothesized that the higher degree of suburbanization in the Randstad may cause this difference.

From in-depth interviews with planning officials in the Ruhr area it was learned that the northern part of the Ruhr area is quite different from the more affluent Cologne–Dusseldorf region. The northern part is more economically depressed due to the decline of the coal pits and the steel industry. The more local orientation and the higher rates of unemployment result in lower levels of mobility, as was shown by secondary analysis of the KONTIV data.

Given the focus of the study on congestion, the portion of the travel taking place in peak periods was analysed. Table 2.8 presents the share of the distance travelled in the four busiest hours of the day. It can be seen that the

*Table 2.8 Share of distance travelled in peak periods by mode of transport**

Mode of transport	Randstad (%)	Ruhr area (%)
Car driver	27	35
Car passenger	16	16
Public transport	41	32
Bicycle	31	24
Walk	14	21
Other	27	13
All trips	27	30

Note: *The peak periods are defined as the four busiest hours in each of the areas. Distance travelled refers to parts of the trips that are taking place in the four hours.

concentration of car travel in the peak periods is much higher in the Ruhr area than in the Randstad. Although the mobility level in the Ruhr area is lower than in the Randstad, car kilometrage in the peak periods is equal in both areas. Given a different composition of travel by purpose, more trips are made in off-peak periods in the Randstad.

Characteristics of Population, Labour Market and Housing Market

The significant differences in mobility patterns between the Randstad and the Ruhr area made an investigation into possible explanatory factors necessary. It was beyond the scope of the study to develop and test a comprehensive model followed by rigorous statistical validation. Therefore an exploratory one-dimensional empirical analysis of some factors has been carried out.

The distribution of the population by age and occupational status was regarded as an important factor. To determine the explanatory power of this factor, differences in this dimension have been established and the effect of these differences on total mobility by mode and by trip purpose have been estimated using category-specific trip rates. The analysis showed that the share of younger people and full-time and part-time workers is higher in the Randstad. The Ruhr area has a much higher percentage of elderly inhabitants. The higher level of home-to-work mobility as well as the higher level of car use can be explained by this difference in population composition. However, the difference in social and leisure trip rates could not be explained despite the use of extensive activity and time budget studies available for both regions.

The second potential explanatory factor is the labour market. Both areas differ in terms of the demand for labour and the regulation of the labour market. The number of jobs in the Randstad has been increasing during the last two decades, whereas the number of jobs has remained stagnant in the Ruhr area. It is well known that, in general, new jobs as well as a change of job lead to more travel. Therefore, a dynamic labour market induces more travel than a rigid and stable market. Another difference is the economic structure. Although a quantitative analysis appeared to be difficult, it is clear that whereas the percentage of jobs in the service sector in the Randstad is about 80 per cent, it much lower in the Ruhr area.

The last difference to be mentioned in connection with the labour market is the degree of flexibility. A comparison of a large number of regulatory aspects showed a general similarity, but a marked difference in flexibility. The Dutch labour market, in particular, allows a quick response to business cycles through flexible temporary contracts as well as layoff regulations. This may imply a higher level of work travel in the Randstad. In terms of the housing market, it was found that home ownership rates are much higher in the

Randstad than in the Ruhr area. It may be hypothesized that this factor suggests a lower number of household moves and a concomitant increase in the work mobility level.

To summarize, the Randstad is a much more dynamic area than the Ruhr area, with higher population growth (8 versus 2 per cent), more employment growth (27 versus 3 per cent) and more housing stock growth (42 versus 28 per cent) between 1970 and 1990.

Spatial Characteristics and Development

A close relationship exists between the spatial layout of activities and travel patterns. Therefore the land-use patterns of the three areas have been compared with respect to population and employment.

From Table 2.9 it can be seen that, compared with the Ruhr area and the ABG area, the Randstad has the smallest concentration of both population and employment in the central districts of the large cities. At the same time, the Randstad has the highest shares in the peripheral districts of these cities. In the same table, the development in these spatial distributions is shown. Growth in population in the Randstad has taken place, in particular, in the smaller cities and the rural areas; the increase in employment has, in addition, occurred in the outer districts and suburban areas of the larger cities. This outward shift has clear implications for the market viability of traditional public transport. In the more stagnant Ruhr area, growth in population and employment occurred only in the rural areas.

Of course, these outcomes are dependent on the definition and demarcation of these types of districts. In order to eliminate this influence, the distribution of population and employment by the distance to the centre of the nearest large city was determined for the Randstad and the Ruhr area. Table 2.10 shows that a significantly larger proportion of the population and employment opportunities is located within 10 km of the city centres in the Ruhr area than in the Randstad. A very strong outward shift of population and employment during the last two decades can be seen in the Randstad. A similar process, but at a much lower rate, has also been taking place in the Ruhr area. From interviews held among local officials it was learned that new residential and office developments could be built within existing built-up areas in the Ruhr area. In the Randstad the spatial constraints did not allow such a policy and these developments had to be planned in peripheral locations.

It can be concluded from additional figures not shown here that because both population and employment shifted outwards simultaneously, the job surpluses/deficits per sub-area were smaller in the Randstad than in the Ruhr area. It should be noted that it remains true that the potential use of public transport is negatively affected by the high level of dispersion.

Table 2.9 Distribution of population and employment among type of district

	Randstad		Ruhr area		ABG area	
	Share 1990 (%)	Increase 1970–90 (%)	Share 1990 (%)	Increase 1970–90 (%)	Share 1990 (%)	Increase 1970–90 (%)
Population						
central areas, large cities	4	–14	8	–12	5	–15
outer districts, large cities and suburban cities	46	–1	43	0	39	+9
small cities	20	+23	27	–2	7	–3
rural area	30	+13	22	+20	48	–2
all districts	100	+8	100	+2	100	+2
Employment						
central areas, large cities	14	–12	17	–6	20	data not available
outer districts, large cities and suburban cities	44	+26	38	0	40	
small cities	19	+29	26	–3	8	
rural area	23	+52	19	+39	32	
all districts	100	+24	100	+3	100	

Table 2.10 Distribution of population and employment by distance to city centre of nearest large city (1990) and growth in the period 1970–90

1990	Population		Jobs	
Distance (km)	Randstad (%)	Ruhr area (%)	Randstad (%)	Ruhr area (%)
0–5	24	21	31	30
5–10	20	32	19	29
10–20	36	32	33	27
20–30	17	10	16	9
30+	3	5	2	4
Total	100	100	100	100
1970–90	**Population**		**Jobs**	
Distance (km)	Randstad (%)	Ruhr area (%)	Randstad (%)	Ruhr area (%)
0–5	–25	–8	–1	–6
5–10	27	–2	36	–3
10–20	26	9	54	12
20–30	16	9	40	15
30+	46	12	43	23
Total	8	2	27	3

7. INFRASTRUCTURE INVESTMENT PROGRAMMES

One of the projects carried out in this research focused on the investment programmes and on the process of decision making in infrastructure planning for the three areas. The study was limited to passenger transport only.

The investment programmes were analysed in terms of budgets, policy priorities and major projects, as well as in regard to their underlying principles from a transport policy point of view. The reason for this analysis was the concern that not only are the transport problems severe in the Randstad, but investment levels are too low, which exacerbates the negative economic effects of congestion.

Another concern was that the time required to reach an investment decision and start implementation is extremely long in the Netherlands. Therefore, the formal procedures and the practice of decision making have been compared in the three areas. Table 2.11 summarizes the major outcomes of the comparison of the investment programmes.

Table 2.11 Characteristics of current investment programmes for passenger transport infrastructure

Item	Randstad	Ruhr area	ABG area
Annual budget (million Dfl.)	2100	2700	1000
Share of rail infrastructure investments (%)	Circa 55	Circa 55	Circa 80
Type of road investment projects	Missing links and some widenings	Missing links and some widenings	Missing links and some widenings
Type of rail investment projects	Internal and external	Mainly external	External TGV
Multimodal investment programmes	Yes	Yes	No
Share of budget from central government (%)	80–90	80–90	80–90 (= Flanders)
Explicit programme evaluation	No	No	No
Explicit project evaluation method	Yes, partially systematized	Yes, standardized	?
Explicit procedure for incorporation of projects	No	Yes	No

It has been found that the annual investment budget of 2.1 billion Dfl. in current programmes is high compared to that of both other regions, given the population size. It is striking that in every region most of the investments are related to rail infrastructure projects. Apart from a few exceptions, no new motorway links are planned. The exceptions all relate to obvious missing links in a mature network. Most road investments are related to systematic upgrading of existing motorways to six or eight lanes. The latter is standard policy in the Ruhr area and the ABG region.

To a large extent rail investments have to do with the external links of the three areas which are being constructed or planned as high-speed-train connections. These projects will require large sums of money for the coming years, limiting the investment options for other purposes. Only in the Randstad are internal rail connections being built or upgraded. Some underground metropolitan rail links will be built in Amsterdam and Rotterdam. This financial infusion may be caused by the need to compensate for lack of investment in such systems during the last decades.

Although the administration is very different in the three areas, in every area the vast majority of the investment money is furnished by the central government. Lower levels of government do not appear to have significant means and plans for strategic investments.

The observation was made that no clear links at programme level could be discerned between the programmes and the policy goals of the study areas. Of course, at an abstract level, references are made to those goals and to visions addressing spatial and economic policy. However, it is difficult to determine the arguments for the choice of particular investment packages. At an abstract level, the underlying economic reasoning is similar in the Randstad and the ABG area: enhancement of the transport sector, including the main ports. In the Ruhr area the primary objective is different: accessibility should serve the service sector and the high-tech industry, including research and development activities.

8. CONCLUSIONS

The traffic volumes of the road network in the Randstad are substantially higher than in the Ruhr area and the ABG area. This disparity is partly caused by the lower supply of road capacity as expressed as a measure of lane length per capita. The comparison showed that, in the Randstad, not only is the per capita motorway capacity lower, but that an underlying main surface road network is largely lacking. The implication of this finding is that, at the motorway level, no alternative routes are available and, furthermore, in case of congestion and incidents on the motorways, few surface diversion options are possible.

It was found that the lack of non-motorway main roads in the Randstad does not lead to an increased use of motorways in short trips. The motorways are being used for many short trips in the Ruhr area simply because the share of short trips is high in that area.

Higher traffic volumes do not necessarily lead to a significantly higher level of congestion. Empirical analysis showed that in the Randstad car trips are distributed more evenly over the hours of the day. The typical inhabitant of the Randstad travels, on average, more than the typical inhabitant of the Ruhr area. However, the Randstad citizen travels to a higher degree by non-car modes and during off-peak hours.

The higher level of mobility in the Randstad in comparison to the Ruhr area is caused, in part, by differences in demographic and socio-economic characteristics of the population. In addition, the Ruhr area is a unique region, which is dominated by the 'Kohlenpott', a district characterized by coal and steel production with little economic or demographic growth.

NOTES

1. This is not shown in this chapter.
2. As car driver.
3. The Flemish data were not available for secondary analysis.
4. Note that no data were available for the ABG area.
5. It should be noted that the travel rates are partially affected by differences in the survey design; response in the KONTIV survey is significantly higher than in the Dutch OVG survey.

REFERENCES

Hilbers, H.D., Wilmink, I.R. and Verroen, E.J. (1996), *Vergelijking aanbod en gebruik hoofdwegennet in enkele Europese metropolen*. Report number: INRO–VVG 1996-04, TNO–Inro. (Comparison of supply and use of the main road network in three European metropolitan areas; in Dutch.)

Hilbers, H.D., Wilmink, I.R. and Krolis, H.P. (1997), *Infrastructuur en mobiliteit in de Randstad, het Ruhrgebied en de regio Antwerpen–Brussel–Gent*. Report number: Inro–VVG 1997-12, TNO–Inro. (Infrastructure and mobility in Randstad Holland, the Rhine–Ruhr Area and the region Antwerp–Brussels–Ghent; in Dutch.)

Hendriks, F., Jansen, G.R.M., Prak, P., Tops, P.W., Zouridis, S. and Esselbrugge, M. (1997), *Infrastructuur investeringsbeleid in vergelijkend perspectief: een bestuurskundige en een verkeerskundige analyse van het Ruhrgebied, de Vlaamse stedenruit en de Randstad*, Tilburg, Catholic University Brabant and TNO-Inro, Delft. (Infrastructure investment policy in comparative perspective: a policy and transport analysis of the Rhine-Ruhr Area, the region Antwerp-Brussels-Ghent and Randstad Holland; in Dutch).

3. The effects of parking pricing and supply on travel patterns to a major business district

Yoram Shiftan

1. INTRODUCTION

It has long been recognized that transportation supply cannot meet demand in city centers. To solve transportation congestion, air pollution and safety problems, transportation control measures (TCM) must be implemented. Central Business District (CBD) auto-restraint policies should be part of any efficient strategy to reduce the use of private cars in city centers. Such policies include tolls, area permits or licenses, physical restrictions, high-occupancy-vehicle priority schemes, a complete ban on cars in selected streets, and parking management. In many countries, governments are increasingly recognizing the use of parking policies as a means of reducing urban road traffic (Barde and Button, 1990; Verhoef et al., 1996; Voith, 1998), and many researchers believe that parking measures are effective means of managing travel demand (McShane and Meyer, 1982; Brade and Button, 1990).

Parking management can be applied on many dimensions, including the number of parking spaces and their spatial distribution, parking rates, time limits, residential parking permits, taxes, employee parking, and level of enforcement. The total amount of parking available in the city center can affect the amount of traffic entering the area, and the location and layout of these spaces can affect the movement of traffic within the center. On-street parking reduces the traffic capacity of roads in and approaching the center. Parking programs, however, do not affect through traffic; they can actually increase it, and they may also increase the number of chauffeur-driven cars.

Parking management can be used to encourage people to shift from private car to public transportation. Parking policy, however, can have other effects on travel patterns. It may encourage people to go to other destinations, change the time of day of the trip, and change or cancel their activities. In the long run, it may even cause businesses to move outside the business district, an outcome that will only serve to disperse activities and further dependence on private vehicles. Such a response to a new parking policy may result in increased congestion and air pollution, thus achieving the reverse effect of the one intended by implementing the measure. Moreover, it is difficult for city centers to compete with shopping malls and office parks that provide free parking. American downtown areas, in consequence, have lost much of their vitality. Between 1976 and 1986, retail centers shifted dramatically away from downtowns of all sizes; today, downtowns are primary shopping areas in only 21 percent of all communities with a population of 50 000 to 500 000 persons (The Urban Institute, 1989).

Given the practical and political problems of collecting tolls, parking fees may be thought to be a good substitute for tolls (Glazer and Niskanen, 1992; Verhoef et al., 1995). Glazer and Niskanen, however, conclude that the cost of parking may be equivalent to an increase in the cost of driving for commuters who park for a fixed length of time, but not for through drivers or for persons who can vary the length of time they park. When these latter types of drivers constitute a significant fraction of all traffic, an increase in the parking fee may have no effect on congestion.

The objective of a parking-management program should be to make the Central Business District (CBD) as attractive as possible by encouraging people to change their choice of travel mode and travel time without discouraging them from coming to the city center. A good parking policy should restrain commuters without hindering shoppers and people doing personal business; it should ensure that commuters do not fill all available parking space and that some unoccupied space is spread over the area so that shoppers and other visitors are not discouraged from visiting the center.

Potential parking schemes to achieve these objectives include varying parking rates by time of day, charging higher rates for longer periods, and limiting parking time. Parking measures should be imposed only during congested peak periods so as not to discourage shopping and other types of midday trips that are important for the vitality of the downtown.

The cost and benefits of parking management programs should be carefully studied before implementation. Benefits include reduced travel time and operating costs for some users, improved downtown amenity and potentially improved economic activity, reduced air pollution, noise and energy consumption, more productive land use, and a reduced need to expand highways. Costs include a potential decline in economic activity, more administration costs, increased transit deficit, increased travel time and operating costs for users who change their travel patterns to avoid the restraint penalty, and potentially increased congestion outside the area where the measure is applied. Finally, revenue from increased parking rates should be considered neither as a benefit nor as a cost, as it is really a transfer of resources.

The question of the effect on economic activity of the downtown is key in any decision to implement parking policies. Parking measures are often presumed to increase the cost of doing business in the downtown area by making the auto trip more expensive and inconvenient. On the other hand, the reduced congestion resulting from such measures will lessen travel time, therefore reducing the price of doing business. The net effect of these outcomes is hard to predict.

To evaluate these benefits, we need a good understanding of people's responses to such measures. We need to understand how parking policies affect the demand for and supply of parking, how parking demand and supply affect the vitality and value of the CBD, and how parking policy acts as a powerful transportation planning tool. Little data and information, however, are available either on the potential impact of parking measures on people's travel behavior patterns or on the long-term effects of such measures. Studies of the effect of parking policies on auto use and congestion in city centers are either empirical studies looking at the before-and-after implementation of a parking policy or simulation studies in which travel demand models are used to evaluate the potential effect of a parking policy. Examples of empirical studies are as follows: Thomson (1967); US Department of Transportation (1982); Mehranian et al. (1986); Surber et al. (1984); Willson (1997); Willson and Shoup (1990); and Shoup and Willson (1992). Examples of modeling and simulation modeling studies include Gillen (1977) and Gomez-Ibanes and Fauth (1980).

The main problem of the modeling and simulation studies is their reliability. The ability of the model to correctly predict the response to parking measures is questionable, especially since these models have usually not been calibrated for such purposes, but have been developed mainly for the evaluation of new infrastructure. The advantage of the empirical studies is that they look at actual changes in travel patterns resulting from the implementation of such policies. There are two main problems with empirical studies: first, it is difficult to isolate the effect of a parking measure from other external effects; second, it is very difficult to identify the trips that are affected the most. If most of the trips affected are work trips, then the measure is very successful; however, if most are shopping trips or other visitor trips, then the measure may have a negative effect.

Given the political fragility of parking policies, there is need for much more evidence of the potential effects of such policies. The character of the benefits makes it difficult to rally political support for parking measures, while the costs are likely to evoke both organized and individual opposition (Howitt, 1980). The purpose of this chapter is to study people's responses to different parking policies through a stated-preference survey. The problems of stated preference are well recognized (see, for example, Beaton et al.,

1997; and Bates, 1988). The aim is not to provide an accurate estimate of potential travel changes as a result of a specific parking policy, but to study the behavioral changes of travelers under such policies and to determine how these policies can influence the city center. The importance of parking policies has long been recognized; as a 1965 UK report says, 'Parking policies are an indispensable part of the land use and transport plan for the town centre' (Her Majesty's Stationery Office, 1965), and parking is still viewed as 'the unstudied link between transportation and land use' (Shoup, 1997).

2. THE SURVEY

A stated-preference survey of 200 people arriving at the Carmel Center Central Business District in Haifa was conducted to study their response to changes in parking policies. The Carmel Center is one of the city of Haifa's few business districts. This center is of mixed land use, catering to residential, commercial, services and leisure activities; as such, it serves as a regional business district for the nearby neighborhood as well as constituting one of the major business districts in Haifa. It is a traditional urban CBD, with small shops, services, offices and leisure activities and no major shopping malls. (It does have a small shopping mall that is part of an apartment and hotel complex.)

The survey consisted of four parts. The first part asked details about today's trip: origin of trip; time of arrival at the center; trip purpose; expected duration of visit; and mode of travel.

The second part was for auto users only and had three sections. The first section asked details about today's parking: the time needed to find parking; to walk from parking location to destination; and the type of parking (paid/free and on/off street). The second section included the stated-preference experiment (see section 4). The last section asked the response of auto users to recent parking-regulation changes in Carmel Center.

The third part of the survey asked people who did not come by private car for the reason that they did not use an auto and the conditions under which they would shift to a private car. Respondents were also asked if they had used their car to come to the center before the recent parking changes.

The last part of the survey contained demographic and socio-economic questions in regard to age, marital status, household size, children, auto ownership, number of drivers, and income.

3. THE SAMPLE

The sample had 29 percent work trips, 20 percent shopping trips, 25 percent entertainment trips, and 26 percent errands trips. Of the respondents, 7 percent arrived from the neighborhood of the Carmel Center itself, 36 percent from nearby neighborhoods (short driving distances of up to 10 minutes), 16 percent from the two other major Haifa business districts, and the rest from neighborhoods further away.

In terms of arriving mode, 52 percent of the respondents came by auto, 28 percent by public transportation, 8 percent by taxi, 2 percent by motorcycle, and 10 percent on foot. Among auto users, 61 percent were alone in the car and 39 percent had at least one passenger. Parking search time was almost evenly distributed among the three ranges provided: 0–5 minutes; 5–10 minutes; and more than 10 minutes. Some 47 percent of the auto users said they walked for up to 5 minutes from their parking location to their destination, 39 percent walked for between 5 and 10 minutes, and 14 percent walked for over 11 minutes. A total of 67 percent of the auto users parked for free on the curbside, 11 percent parked in employee parking, 13 percent paid for curb parking and 9 percent paid for garage parking. Most parking was for short duration: 27 percent of the respondents expected to stay in the Carmel Center for less than an hour; 28 percent for less than a couple of hours; and 17 percent for less than three hours. The rest of the sample, 28 percent, expected to stay for more than three hours; 80 percent of this group were people who had come to work.

4. RESULTS

Two parking measures were studied. The first was an increase in hourly parking rates from their current 3.70 New Israeli Shekels (NIS) to 5, 7 and 10 NIS. The second measure was a reduction in the number of available parking spaces. The reduction in parking spaces was presented to the respondents in term of an increase in the time needed to find a parking space of 10, 15 or 20 minutes. For each question, respondents were given the option to cancel their trip, to change destination, to change the time of the trip, to shift to public transportation or to taxi, to walk, or not to change their travel behavior.

Stated Response to Parking Price Increase

Figure 3.1 shows the response to price increases to 5, 7 and 10 NIS. For a fee of 5 NIS per hour, 71 percent of those who arrived at the center would not change their travel behavior. Of the 29 percent who would, 12 percent said they would

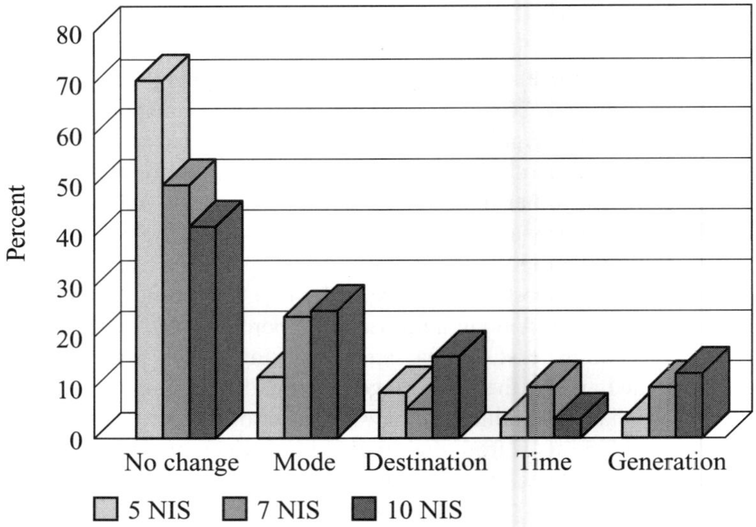

Figure 3.1 Responses to parking price increases, all respondents

change mode, 9 percent would change destination, 4 percent would change the time, and 4 percent would cancel their trip. Increasing the price to 7 NIS would cause 50 percent to change their behavior, and a fee of 10 NIS would cause 58 percent to change their behavior. Although these results are interesting, it is more important to look at them by trip purpose. Figure 3.2 shows the results for work trips. Most of the commuters who would change their travel behavior as a result of a parking price increase would shift mode, and some would shift the time. There is no destination change or elimination of trips for work trips. For non-work trips (Figure 3.3), all types of changes occur, the most common being mode shift and destination change, both at about the same rate.

A good parking policy should encourage workers to shift mode and time of day, but not discourage visitors from coming to the center. For purposes of analysis, therefore, all travel-behavior responses were categorized into three groups:

- no change in behavior;
- change in mode or time of day – this is considered a positive change, as it will reduce traffic without reducing the number of visitors to the center;
- change in destination or cancellation of the trip – this is considered a negative change, as it will reduce the number of visitors to the center and, therefore, may affect the vitality of the center.

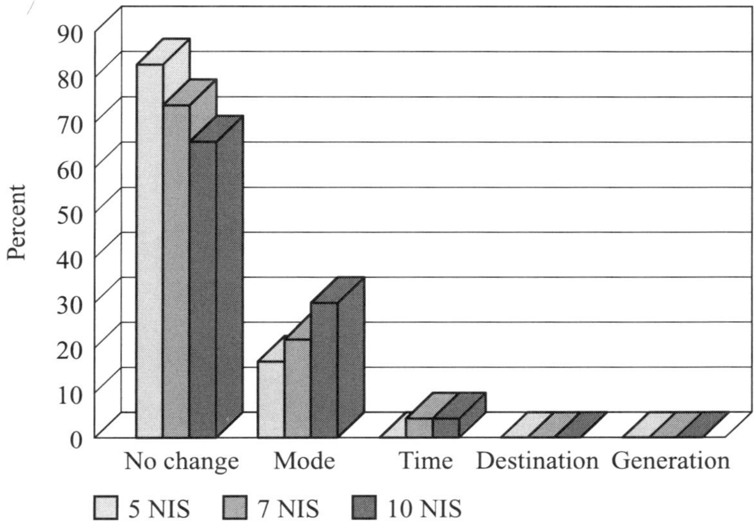

Figure 3.2 Responses to parking price increases for work trips

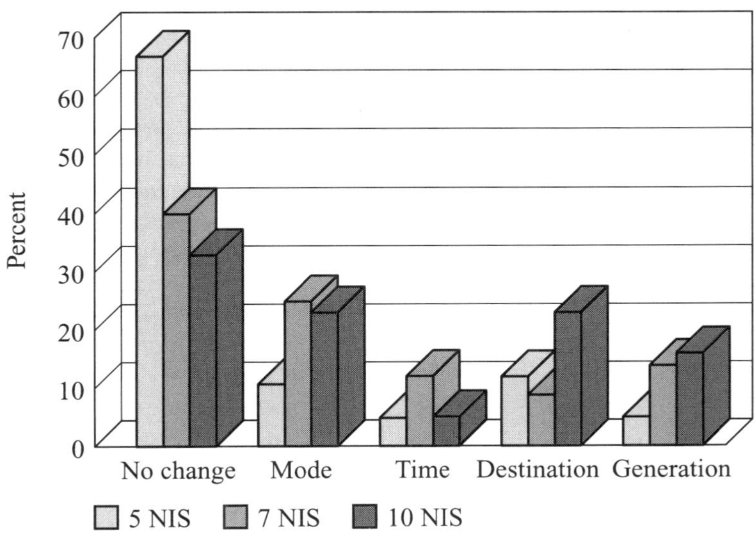

Figure 3.3 Responses to parking price increases for non-work trips

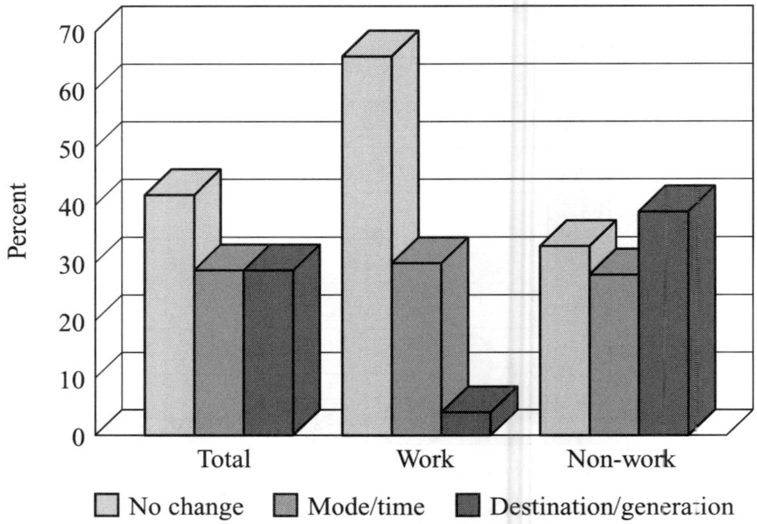

Figure 3.4 *Responses to parking price increases to 10 NIS for work and non-work trips*

Figure 3.4 shows the response by these groups to a price increase to 10 NIS for both work and non-work trips. For the whole sample, there is an equal number of people who state they would change mode or time and who state they would change destination or cancel their trip. A look at these results by trip purpose shows that among those on a work trip, 30 percent would change mode or time and only 4 percent would change destination or frequency. This result suggests that the policy can be a good one for work trips. The proportion of non-work travelers who state that they would shift mode or time of day is similar (28 percent) to that of commuters; however, there is an additional 39 percent who stated they would either change their destination or cancel their trip. Such a travel-behavior response can have a significant effect on the vitality of the center and eventually cause businesses to move out of the center. In the long run the policy can indirectly affect work trips as well, leading people to change their destination or cancel their trip.

Stated Response to Parking-availability Measures

Figure 3.5 shows the main results of the response to parking-availability measures. Some 47 percent of the respondents will not change their behavior under a parking-supply measure that results in 10 minutes of parking search

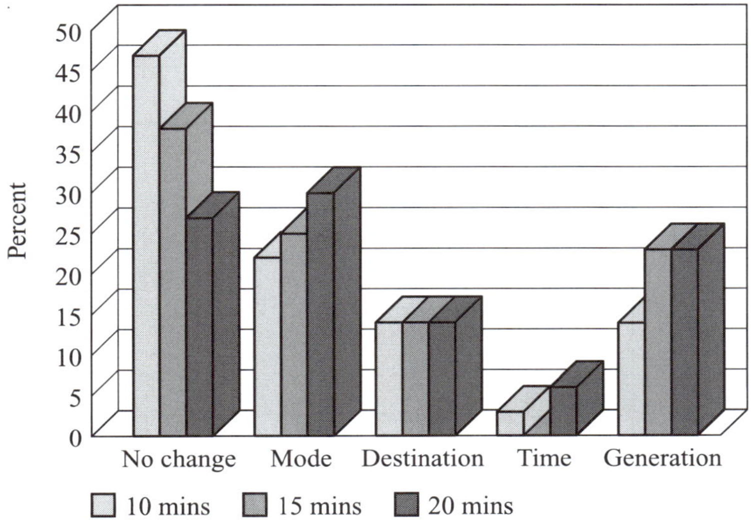

Figure 3.5 Responses to parking-availability measures for all trips

time; the percentage goes down to 38 percent for a 15-minute search time and to 27 percent for a 20-minute search time. The dominant travel behavior change is shifting mode, followed by reducing the frequency of trips, then by shifting destination, and lastly by changing the time of day.

Figure 3.6 shows the results for work trips. The parking-availability measure produces a stronger effect than does the pricing measure on mode shift for the tested range. With a parking search time of 10 minutes, 35 percent of the work trips will shift to transit. A further increase in parking search time, however, has only little marginal effect on mode shift; and with a parking search time of 20 minutes, 39 percent will shift to transit. There was no destination or time-of-day shift for work trips; and a small percentage (4 percent) stated they would cancel their trip for all three levels of the measure.

Figure 3.7 shows the results for non-work trips. For a 10-minute parking search time, 18 percent will shift mode, 19 percent will shift destination, 4 percent will change time of day, 18 percent will cancel their trip, and 42 percent will not change their behavior. Although the percentage of changed behavior increases as the parking search time increases, and mode shift also increases, destination change does not increase beyond its 19 percent level even for higher parking search times. The percentage canceling their trip increases from 18 percent for the 10-minute parking search time to 30 percent for the 15-minute parking search time, and it remains 30 percent for the 20-minute parking-search time. The time-of-day shift is limited to a small

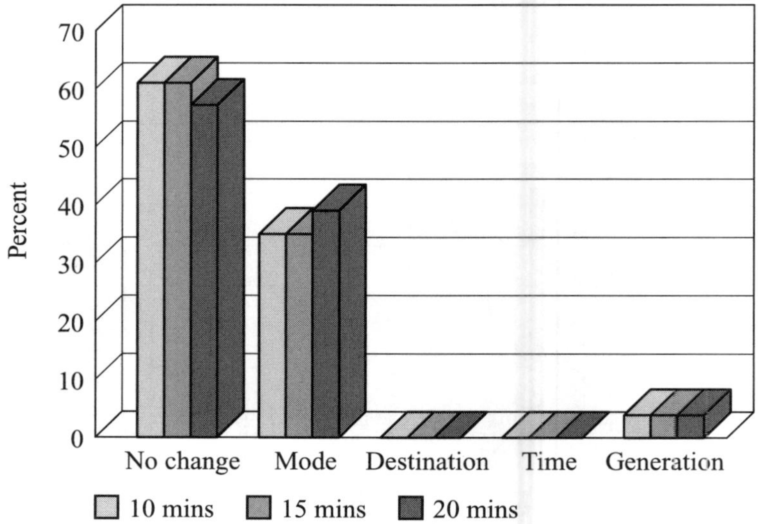

Figure 3.6 Responses to parking-availability measures for work trips

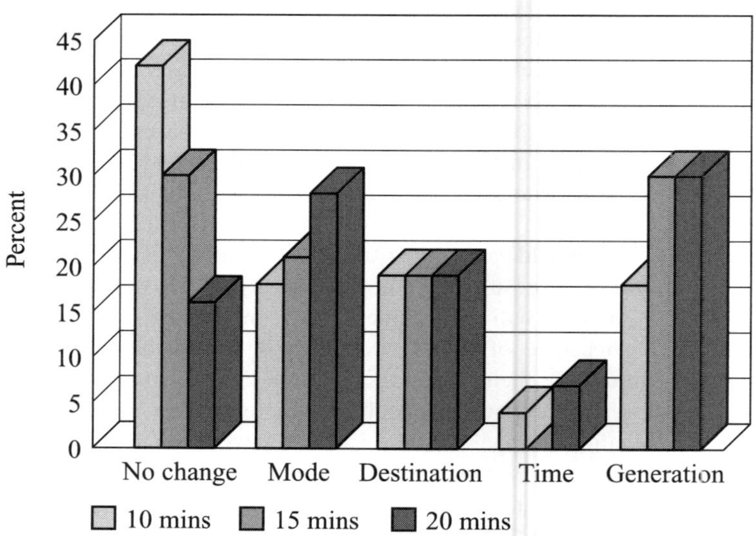

Figure 3.7 Responses to parking-availability measures for non-work trips

percentage, varying from 4 percent for the 10 minutes of parking search time to zero percent for the 15-minute parking search time and 7 percent for the 20-minute parking search time.

Figure 3.8 summarizes the positive and negative responses for the 20-minute parking search time for work and non-work trips. All trips have similar positive responses: 39 percent of the work trips will shift mode or time of day and 35 percent of the non-work trips will make such shifts. There is only a negligible negative response for work trips (of 4 percent), whereas there is a significant negative response for non-work trips of 49 percent. This result has significant policy implications. While work trips will continue to arrive at the center, the exclusion of non-work trips will gradually cause the center to deteriorate and in the long run may cause businesses to leave the center. In the long run, too, work trips will also shift destination, and this shift may even result in a shift from transit to auto mode if the new location is not served well by transit.

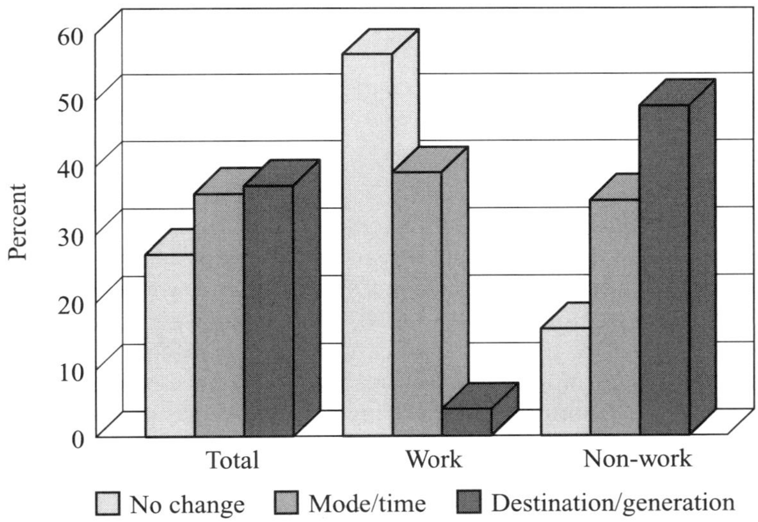

Figure 3.8 Responses to 20-minute parking search time for work and non-work trips

Response to Parking Supply Versus Parking Charges

Figure 3.9 compares the response of both work trips and non-work trips to parking-supply measures versus parking-charges measures. As expected, non-work trips are more flexible and, therefore, more likely to change their

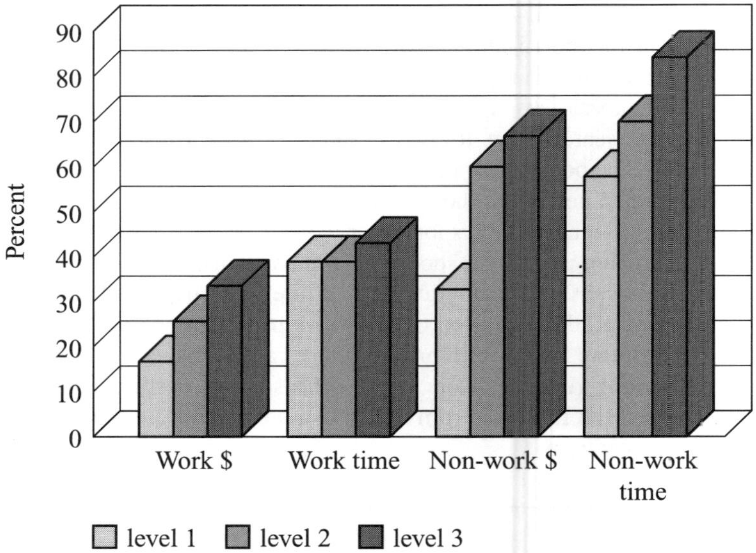

Figure 3.9 Responses of auto users making work and non-work trips to parking-availability measures vs parking price increase measures

behavior. As discussed above, this higher percentage results from a tendency to change destination and to cancel the trip in addition to change mode and time of day, which occur at a similar rate for work and non-work trips. For the range of measures tested, the availability measure shows a higher effect than the pricing measure. Although this result is a function of the defined ranges of the two measures, the importance of parking availability in comparison to parking pricing can also be seen from two other questions.

The first question was asked to transit and taxi users: why were they not using a private auto? A total of 66 percent indicated they had no driving license or vehicle. Of the remaining respondents, 19 percent pointed to parking availability and only 6 percent answered because of parking prices. A further 9 percent answered because transit was more convenient, which may also be related to parking availability. Although these numbers are not significant, they do indicate that parking availability is a more serious problem than parking prices.

In the second question transit users were asked what conditions would cause them to shift to a private auto. In response, 55 percent of transit users indicated they would never shift to an auto, 25 percent that they would shift if parking were more readily available, 15 percent that they would use an auto only in emergency circumstances. Only 5 percent replied that they would use an auto if either congestion were reduced or parking prices were lower. The responses to this question show, as well, that parking availability is of more concern than are parking prices.

Revealed Response to Recent Parking Regulations

Recent parking-regulation changes in the Carmel Center included the conversion of curb-free parking spaces to paid parking for a limited duration of one or two hours between the hours of 9 am and 8 pm. Auto users were asked if their travel patterns had changed as a result of these measures. Possible answers: changed frequency of arrival, changed time of day, or underwent no change in travel patterns.

Figure 3.10 presents the response of auto users to these recent parking regulations. The results show that, overall, 46 percent of the drivers changed their travel behavior as a result of these changes. Among work trips, 21 percent changed their behavior; and among non-work trips, 56 percent changed their behavior. The majority of the change was manifested in reducing the frequency of visits, with some changes recorded in time of day.

These results deserve some discussion. The results for work trips are clear. Due to the measure there is less curb parking available, and therefore there is

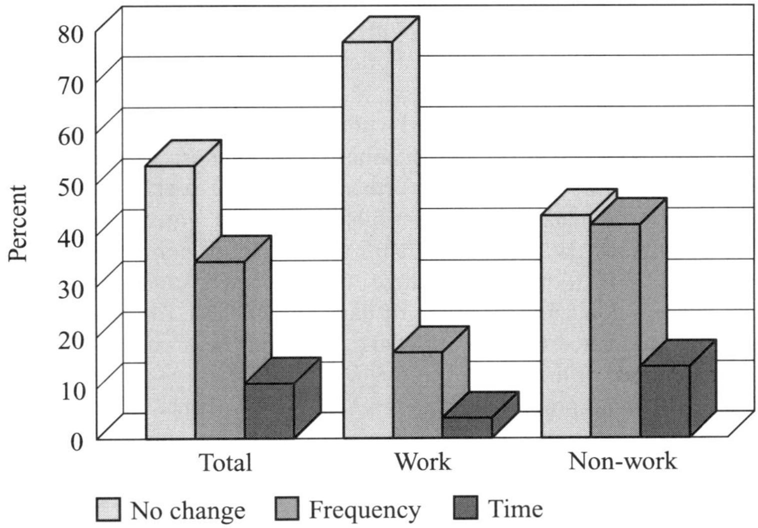

Figure 3.10 Response of auto users to recent parking regulations

a change in work-trip behavior. This change is limited, given the low flexibility of these trips.

The significant change for non-work trips raises some questions. As a result of the change in parking regulations, we would expect more space to be available for short-term parking. The above analysis suggests that parking availability is a more significant concern than are parking charges: 55 percent of the respondents came to the center for less than two hours and should benefit from more available parking, though at a higher cost as a result of the measure. This suggests an expected increase in non-work trips. One explanation for this unexpected result may be bias response. People do not like to pay for parking and demonstrate their objection to an increase in parking prices by stating that they come less often because of this charge.

Transit users were asked about their mode of travel before the change in regulations. Among work trips, 21 percent used an auto before the change; and among non-work trips, 35 percent. These results raise the same question as for the auto users. Work trips are less flexible and, therefore, may change less. On the other hand, non-work trips may benefit from the measures and, therefore, should change less.

5. CONCLUSIONS

The main results of this study show that the introduction of parking measures will cause people who travel to work by car either to continue to travel by car or to shift to public transportation. With a price of 5 NIS, 17 percent of the work trips will shift to transit; with a price of 7 NIS, 22 percent will shift to transit; and with a price of 10 NIS, 30 percent will shift to transit. The parking-availability measure showed a stronger effect on mode shift for the range tested. With a parking search time of 10 minutes, 35 percent of the work trips will shift to transit. A further increase in parking search time, however, has little marginal effect on mode shift; and with a parking search time of 20 minutes, 39 percent will shift to transit. For non-work trips there was a smaller shift to transit, but there was a larger destination shift, more activity cancellation, and some time-of-day shift. Overall, for non-work trips, more people indicated they would change destination or cancel the activity than change mode or time.

Clearly, the main problem of this research is its reliance on stated-preference data. Parking studies are complicated. Reliance on revealed-preference data has its own problems: mainly, it is difficult to isolate the effect of a parking measure from other external effects. It is also very difficult to identify which trip purposes are affected. If most of the affected trips are work trips, then the measure is very successful; however, if most of the affected trips are shopping or other visitors, then the measure may have a negative effect. Further research is needed to combine revealed and stated preferences in parking studies, using available methodologies for this purpose (see, for example, Ben-Akiva and Morikawa, 1990; and Bradley and Kroes, 1990). Although the magnitude of these results is questionable because of their reliance on stated preference, the purpose of this study was to test the different effect of such measures on work and non-work trips as a general trend, not in exact figures.

The results suggest that such measures may be effective in reducing congestion in the business district; however, they may also have a negative effect on the vitality of the business district. Their effect on regional travel patterns and air quality is not yet clear. Further research and detailed local studies are required before such measures can be implemented.

The planning implications of these conclusions in regard to parking policy are as follows:

- visitors and shoppers should be given priority by fixing appropriate time limits and charges;
- preference should be given to short-term parking;
- on-street parking charges should result in at least some further availability of parking spaces for short-term parking.

ACKNOWLEDGMENT

The contents of this chapter are based on research sponsored by the Henry Ford II Transportation Research Fund. The opinions and conclusions expressed or implied in the chapter, however, are those of the author. The author would like to thank the Henry Ford II Transportation Research Fund for their support of this project.

REFERENCES

Barde, J.-P. and K. Button (eds) (1990), *Transport Policy and the Environment: Six Case Studies*, London: Earthscan.

Bates, J. (1988), 'Econometric Issues in Stated-Preference Analysis', *Journal of Transport Economics and Policy*, **22** (1).

Beaton, P., Q. Chen and H. Meghdir (1997), 'Stated Choice for TDM Models: A Study in Predictive Validity', paper presented at the 76th Annual Meeting of the Transportation Research Board, Washington, DC.

Ben-Akiva, M. and T. Morikawa (1990), 'Estimation of Switching Models from Revealed Preference and Stated Intentions', *Transportation Research A*, **24**.

Bradley, M.A. and A.J. Kroes (1990), 'Simultaneous Analysis of Stated Preference and Revealed Preference Information', *Proceedings of the 18th PTRC Annual Summer Meeting; Seminar H*, University of Sussex, UK.
Gillen, D.W. (1977), 'Estimating and Specification of the Effects of Parking Costs on Urban Transport Mode Choice', *Journal of Urban Economics*, **4**, 186–99.
Glazer, A. and E. Niskanen (1992), 'Parking Fees and Congestion', *Regional Science and Urban Economics*, **22**, 123–32.
Gomez-Ibanez, J.A. and G.R. Fauth (1980), 'Downtown Auto Restraint Policies', *Journal of Transport Economics and Policy*, May.
Her Majesty's Stationery Office (1965), 'Parking in Town Centres', *Planning Bulletin 7*, Ministry of Housing and Local Government, Ministry of Transport, Scottish Development Department, The Welsh Office.
Howitt, A.M. (1980), 'Downtown Auto Restraint Policies: Adopting and Implementing Urban Transport Innovations', *Journal of Transport Economics and Policy*, May, 155–67.
McShane, M. and M.D. Meyer (1982), 'Parking Policy and Urban Goals: Linking Strategy to Needs', *Transportation*, **11**, 131–52.
Mehranian M., M. Wachs, D. Shoup and R. Platkin (1986), 'Parking Cost and Mode Choices Among Downtown Workers: A Case Study', paper presented at the 1987 Transportation Research Board Meeting, Washington, DC.
The Urban Institute (1989), 'Revitalizing Downtown Areas: Guidelines for Success', *The Urban Institute/Policy and Research Report*, Winter and Spring.
Thomson, J.M. (1967), 'An Evaluation of Two Proposals for Traffic Restraint in Central London', *Journal of the Royal Statistical Society*, Series A, 130.
Shoup, D. (1997), 'The High Cost of Free Parking', *Access*, Spring (10), 2–9.
Shoup, D. and R. Willson (1992), 'Employer Paid Parking: The Problem and Proposed Solutions', *Transportation Quarterly*, **46**, April, 169–92.
Surber, M., D. Shoup and M. Wachs (1984), 'Effects of Ending Employer-Paid Parking for Solo Drivers', *Transportation Research Record*, **957**, 67–71.
US Department of Transportation (1982), *Innovations in Parking Management*.
Verhoef, E.T., P. Nijkamp and P. Rietveld (1995), 'The Economics of Regulatory Parking Policies: The (Im-)possibilities of Parking Policies in Traffic Regulation', *Transportation Research A*, **29**, 459–64.
Verhoef, E.T., P. Nijkamp and P. Rietveld (1996), 'Regulatory Parking Policies at the Firm Level', *Environment and Planning C: Government and Policy*, **14**, 385–406.
Voith, R. (1998), 'The Downtown Parking Syndrome: Does Curing the Illness Kill the Patient?', *Business Review*, January–February.
Willson, R.W. (1997), 'Parking Pricing Without Tears: Trip Reduction Programs', *Transportation Quarterly*, **51**(1), Winter, 79–90.
Willson, R.W. and D.C. Shoup (1990), 'Parking Subsidies and Travel Choices: Assessing the Evidence', *Transportation*, **17**, 141–57.

PART II

Telecommunications and Travel

4. Work-related travelling and telecommunications: Substitution, stimulation and complementarity

Mervi Lehto and Veli Himanen

1. INTRODUCTION

For many years, researchers have been forecasting the potential impacts of telecommunications on travelling (for the scientific debate see Plaut, 1997). Because many telecommunication devices are already in use – particularly in the small information-intensive regions such as Scandinavia, Israel, and so on – their impacts on travelling can already be observed. This chapter will assess the impacts of telecommunication on travelling through responses made to a questionnaire among Finnish office workers who use telematic devices in their work. The future potential of transport options, as well as that of telematic devices, is considered according to their current stage in a development life-cycle. The relationship between telecommunications and travelling is also considered here from the viewpoint of current policies as well as from their relative status among human activities. It should be noted that much of the basis for this chapter comes from previous work completed in the projects EMRECU, POSSUM and START, financed by the European Commission.

The chapter aims to assess possible substitution, stimulation and complementarity effects of telecommunications on work-related travelling and to discuss these in the context of sustainable transport policy in Europe. It begins with an examination of human activities and technological developments and continues through a discussion of travelling and telecommunications results reported in a detailed Finnish field survey. The results of this survey are then used to discuss the capabilities of telecommunications to stimulate, substitute and complement work-related travelling. With these data, the potential for using telecommunications for mitigating the negative impacts of travel reductions is considered and conclusions are provided on the relationship between telecommunications and travelling.

2. THE CONNECTION BETWEEN TELECOMMUNICATION AND TRANSPORT

Human Activities and the Limits of Growth

There are plenty of forecasts, especially among futurists, which consider the current era to contain a seed for global revolution (see e.g. Bruckmann, 1987; and King and Schneider, 1991). The main reason for these forecasts is the recognition of both the transition from an industrial to a post-industrial age and the limits of growth towards which the current expansion of human activities is approaching.

When considering the human activities presented in Figure 4.1, it can be seen that basic activities have been the same since the Stone Age and, even though various specialized activities have emerged, several ancient interlinking activities, such as transport, communications and trade, have remained valued. In modern times, these particular activities have gained prestige. However, they are actually only serving the other activities, and their value is related to the specific activities they serve.

Transport is one of the major human activities that consumes non-renewable resources – especially oil – and, at the same time, produces large amounts of environmental pollution. For those reasons many restrictions have been proposed, especially relating to car traffic (see e.g. Dugonjic et al., 1993). On

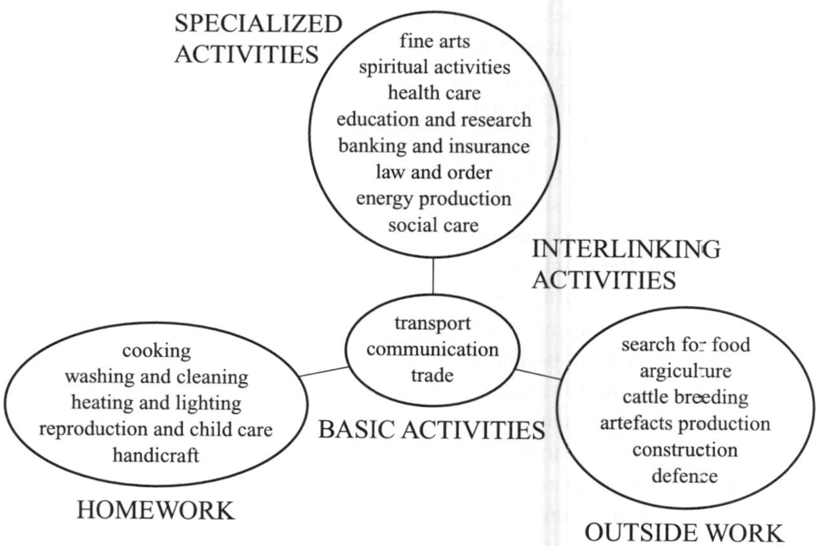

Figure 4.1 Basic, specialized and interlinking human activities

the other hand, telecommunications have generally been considered to include the potential to improve the environment by reducing the demand for travel. However, Marvin (1997) warns of the opposite possibility: that communication activities might generate new demands for travel.

The transition from an industrial to a post-industrial age entails the ever-increasing growth in the use of telematic devices in the post-industrial countries, and the transfer of the production of goods to the countries with cheaper labour forces. The countries with the biggest demand for well-educated labour are the post-industrial or information-intensive ones, where information itself is increasingly an economically significant commodity.

Technological Development

Many human phenomena include trends and fluctuations. The analysis of this has resulted in an interpretative theory of long-wave development, which has been applied to both economic and technological change (see Vasko, 1987; and Ayres, 1990). A similar approach has been applied to industrial products, which, according to management science, go through a life-cycle including a slow introduction, a growth and a maturity phase.

Marchetti (1987) has interpreted past and future changes in the European transport sector by the same method. According to his analyses, the volume of intercity passenger transport by car will be stable and passenger air traffic will grow fast. In other words, transport by car has already reached maturity phase in its life-cycle, but aviation still seems to be in the growth phase.

In the telecommunication sector it is thought that traditional telephony has reached maturity phase, but, in spite of that, new developments in data processing have opened new markets and also the network constructed for telephone connections is now used for telefax, mobile phone and internet connections. Telefax, too, has reached the maturity phase – actually its growth phase was only 10 to 20 years. Devices like mobile phone and e-mail/internet applications are obviously in the growth phase. On the other hand, video-conferencing applications are still in the introduction phase.

Time and Place

Transport and communications are interlinking activities, which by their nature are related and restricted by spatial and temporal dimensions. While transport – moving physical entities – connects different places with limited speed, most communication devices provide a connection between different places without a noticeable time lag.

All telematic devices pose various time and place restrictions to both senders and receivers. Communication situations, depending on the device used, can be divided into two groups: delayed interaction or real-time interaction (see also Figure 4.4). The first group devices may cause fewer time and place restrictions to the senders and receivers. The physical movements needed for meetings pose the most severe temporal and spatial restrictions.

Travelling, especially in connection with work-related activities, is restricted by time and money resources available. It has been noticed that there is a human tendency to expend on average roughly seventy minutes for daily travelling (for a short description of the scientific debate see Himanen et al., 1998). Until today, the increase of car ownership has been the major factor that has made it possible to increase the distance travelled within this seventy-minute limit (see Figure 4.2). However, there are at least some signs of approaching a limit in car ownership (see Figure 4.3).

For work-related activities, it can be assumed that in the long run the mix of physical travel and telecommunication will be optimized in order to obtain maximum possible profit (for economic criteria for contact mode choice see Button and Maggi, 1995). This means optimizing the time and other resources used for travelling, working alone and communicating face to face or with the aid of different devices (see Figure 4.4).

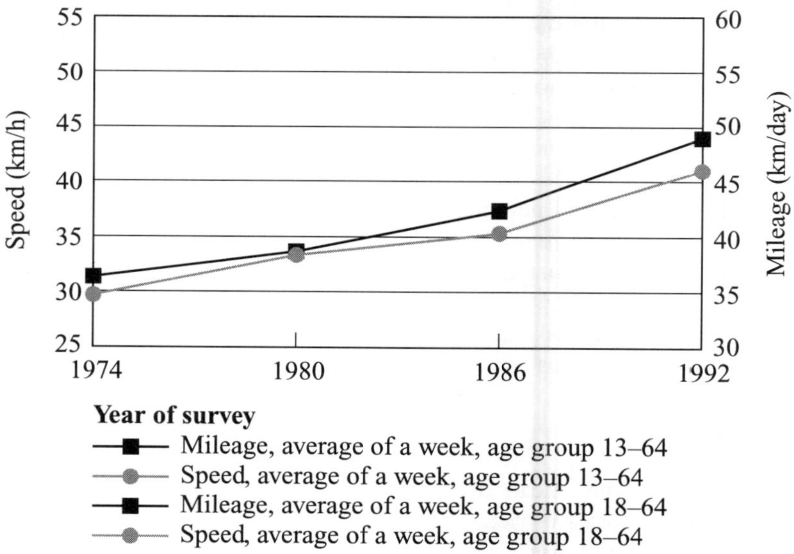

Source: Data from the national travel surveys of the Finnish National Road Administration.

Figure 4.2 Average daily travel speed and distance per person in Finland, 1974–92

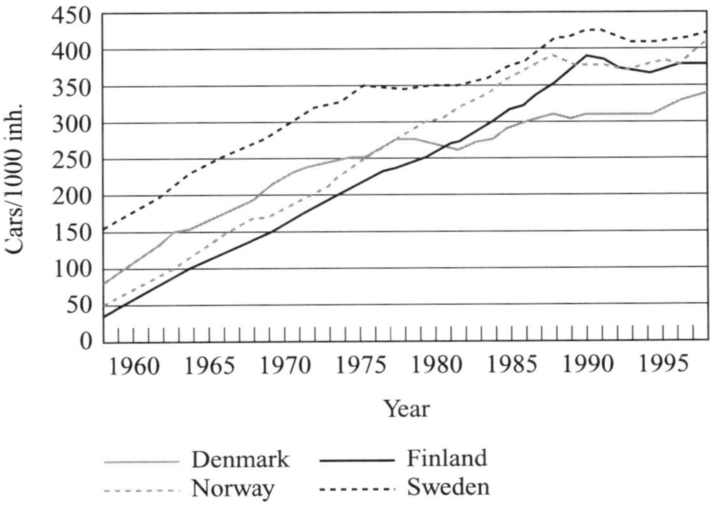

Figure 4.3 *Car ownership in the Nordic countries*

3. SURVEY AMONG OFFICE WORKERS IN THE HELSINKI METROPOLITAN AREA

Method

The results of the effect of telematics on the work-related trips discussed in this chapter were taken from the study: 'Information Transfer Saves Time and Trouble: Reduction of the Use of Energy for Work-related Passenger Transportation' (Himanen et al., 1996) and the 'Intelligent Building' (Lehto, 1996, 1997, 1998a, 1999a; Lehto and Karjalainen, 1997; and Lehto et al., 1997). These studies were undertaken in 12 office buildings surveyed in the Helsinki Metropolitan area, during 1994–95. Five of the offices are built according to the intelligent building concept.

Information was compiled by a questionnaire with a wide scope. It gathered data on the use of telematics and travelling as well as indicators later used in assessing the feasibility of various telematic devices. Over 500 office workers answered the questionnaire to garner a 38.3 and 28.5 per cent response rate respectively in the two phases of the study. The sample is congruent with the characteristics of all Finnish workers, except that education level is higher, half of respondents having an academic degree (see also Lehto, 1996). For the state of the art of the use of information and telecommunication devices in Finland see Appendix 1.

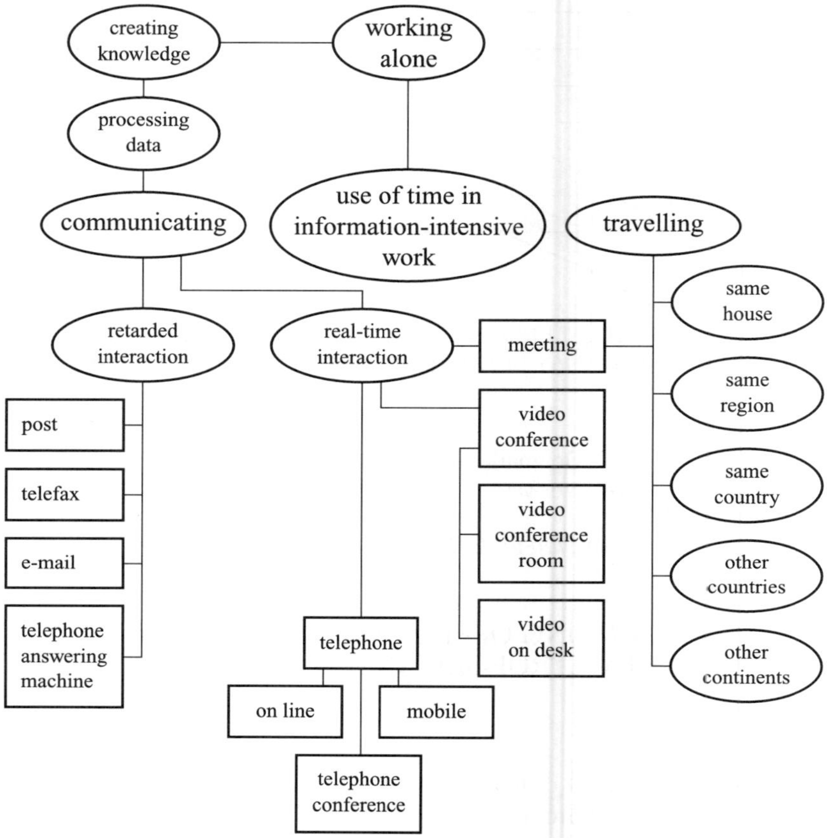

Figure 4.4 *Use of time in information-intensive work and ways of communicating*

Number of Work-related Trips

Regardless of working position and gender, one to two work-related trips[1] per week is a standard for an office worker according to this questionnaire – except for female clerks, who travel less than the other groups (see Table 4.1). The figure is similar to that of the Finnish employees on average, who make 0.28 work-related trips per day (see Himanen et al., 1996).

When comparing male and female supervisors, experts and clerks in the intelligent buildings (IBs) and in the non-IBs, it was found that those who travel most are male supervisors working in the IBs. They make more than two trips per week inside the metropolitan area, almost four trips per month

Table 4.1 *The number of work-related trips to different destinations, according to work position, gender and office building type*

Working position	Supervisor				Expert				Clerk			
Gender	Male		Female		Male		Female		Male		Female	
Office building type	IB	Non-IB	IB	Non-IB	IB	Non-IB	IB	Non-IB	IB	Non-IB	IB	Non-IB
Trips within the metropolitan area per week	2.2	2.3	1.9	2.1	1.8	1.6	1.3	1.5	1.5	0.8	0.6	0.6
Trips to other parts of the country in a month	3.7	1.3	0.4	1.0	2.1	3.1	2.1	1.4	0.4	2.8	0.5	1.2
Trips abroad in a year	7.4	3.6	8.4	0.4	3.5	1.2	2.2	0.6	1.6	0.3	0.8	0.6

Note: IB = Intelligent building.

heading for other parts of Finland and 7.4 trips per year abroad. Also, the female supervisors working in the IBs travel for their work. They have more trips that take them abroad (8.4 per year), but very few trips (0.4 per month) to other parts of the country.

In all, experts and clerks travel far less than supervisors in both the metropolitan area and abroad. Although, according to these statistics, almost everyone makes a trip abroad each year and a few trips per month heading for other parts of the country, personal differences might exaggerate these data. Some individuals or working groups might not travel at all.

4. THE EFFECT OF TELECOMMUNICATIONS ON WORK-RELATED TRAVEL

Substitution

The office workers were asked how many trips they make to different destinations, and if telematics has had an influence on the number of the trips. They were also asked to characterize any influence of telematics in increasing or reducing work-related trips. One-third of the office workers experienced a diminishing effect of telematics on work-related trips. Most of the office workers, almost two-thirds, have experienced no effect of telematics on work-related trips. Extremely few have experienced an increase in travel because of telematics. Also, the share of people who did not have an opinion on this issue was a minor one (Table 4.2). It was more often a man than a woman who reported having experienced the diminishing effect. Female office workers either have not experienced an effect or cannot say. Female office workers make fewer trips than male ones, which might be the cause of these responses.

It is more often a supervisor than a representative of the other working groups who reports no effect of telematic services on the amount of his or her work trips. This observation is particularly valid among female supervisors. The experts have experienced the diminishing effect of telematics on work trips more often than supervisors. However, they share the opinion of no effect with supervisors concerning longer trips. Experts make fewer trips than supervisors, which might have some influence on their opinion. Also the types of trips can vary among experts compared to supervisors, which may give rise to different possibilities for replacing the trips with telematic services.

The results suggest that there is almost no effect of age on the potential of telematics in reducing work-related trips. This finding is parallel to that of the study on home automation, which found no strong correlation between automation and age. In this case, only a few points separate the younger generation from their seniors (see also Lehto, 1998b, 1999b).

The proof of the substitution of work-related trips by telecommunications rests on the opinion of people – how they have experienced substitution in their daily working lives. The magnitude of the substitution remains unanswered here – maybe it can never be defined because of the complexity embedded in the development of working habits. The opinions of the office workers on the substitution cannot be verified in this part of the survey. However, good confirmation of the reliability of the office workers' opinions was found in another part of the survey which compared the measurements of the working environment to the opinions of end users about those measurements (Lehto, 1996). For example, the office workers' opinions on the indoor air temperature correlated to the measured values (Lehto and Karjalainen, 1997). Half of those who answered had a university degree, which may explain this observation of the office workers good judgement.

The third having experienced the diminishing effect of telematics on travelling are largely men with either expert or supervisory positions (see Table 4.2). Those who have experienced the diminishing potential of trips when travelling abroad have the best salary (see Table 4.2) and make a majority of the trips abroad (see Table 4.1). This finding suggests that there is good potential for telematics to save travelling costs when both the travelling costs and the labour costs are high. The grand total of travelling costs is dependent on two factors: the cost per trip and the number of trips made. According to this survey, both factors are influenced by the status and the gender of the employee. Higher working statuses and higher salaries resulted in more and longer trips. If savings in travelling costs due to telematics can be made among those having high status, more money is saved than in other working groups. This often seems to be the case among male supervisors and experts.

The situation is different for women. For women, higher working status means earning higher salaries and making more and longer trips than the other female working groups, but not in comparison to men. Also, women more often than men report no effect of telematics on trips. The savings potential of reducing travelling costs among female office workers is thus, for the time being, not as clearly dependent on working status as among men.

Heavy travellers

The experience of the potential of telematics to affect work-related trips depends on how often a person is travelling. The number of those who do not travel at all for business is very small (4 responses), as is that of those who have experienced the increasing potential of telematics (3–4 responses, 1–2 per cent). Thus the interest is next focused on the other groups (Table 4.3).

Table 4.2 Office workers' experience of the effect of telecommunications on work-related trips to different destinations according to gender, work position, salary and age

Destination/effect (%)	+ increase	– diminish	0 no effect	Cannot say
Within the metropolitan area	1	33	59	7
Other parts of the country	1	27	63	9
Abroad	2	17	69	11

Destination/building type*	IB	Non-IB	IB	Non-IB	IB	Non-IB	IB	Non-IB
Within the metropolitan area	1	1	32	35	60	58	7	6
Other parts of the country	2	1	**29**	23	60	**69**	10	7
Abroad	3	0	18	15	67	**74**	12	11

Destination/gender (%)	Male	Female	Male	Female	Male	Female	Male	Female
Within the metropolitan area	1	2	**38**	27	58	62	3	10
Other parts of the country	2	0	**35**	10	58	**74**	5	15
Abroad	2	0	**22**	7	69	**75**	7	18

Destination/work position (%)	Supervisor	Expert	Clerk	Supervisor	Expert	Clerk	Supervisor	Expert	Clerk	Supervisor	Expert	Clerk
Within the metropolitan area	0	2	2	32	**38**	**22**	**66**	57	61	2	3	15
Other parts of the country	0	2	0	28	30	**15**	**70**	62	65	2	6	21
Abroad	0	2	0	18	16	**8**	**80**	70	67	2	11	25

	Supervisor		Expert		Clerk		Supervisor		Expert		Clerk	
Destination/position and gender (%)	M	F	M	F	M	F	M	F	M	F	M	F
	30	**40**	41	33	38	16	68	60	56	60	50	66
	34	0	36	14	36	6	63	**100**	57	74	50	71
	23	0	18	13	30	0	74	**100**	71	69	50	73
Destination/effect and salary (FIM)												
Within the metropolitan area	13.625 (n=4)				14.202 (n=94)			14.985 (n=165)			**11.767** (n=15)	
Other parts of the country	14.833 (n=3)				15.206 (n=68)			14.552 (n=165)			**12.310** (n=21)	
Abroad	13.667 (n=3)				**18.186** (n=35)			15.204 (n=152)			**11.457** (n=23)	
Destination/effect and age (years)												
Within the metropolitan area	38.5 (n=2)				38.1 (n=44)			**41.1** (n=69)			35.6 (n=7)	
Other parts of the country	44.0 (n=1)				39.6 (n=31)			40.7 (n=77)			38.7 (n=9)	
Abroad	– (n=0)				40.4 (n=15)			**41.0** (n=63)			35.7 (n=9)	

Note: * IB=Intelligent building.

Table 4.3 Office workers' experience of the effect of telecommunications on work-related trips to different destinations according to frequency of work-related trips

Destination: within the metropolitan area – percentage of answers					
Work-related trips	+increase	–diminish	0 no effect	Cannot say	n
Daily	2	**32**	64	2	50
Often	1	36	60	3	118
Occasional	1	31	57	**11**	146
Never	0	25	75	0	4
Destination: other parts of the country – percentage of answers					
Work-related trips	+increase	–diminish	0 no effect	Cannot say	n
Daily	4	**29**	60	7	45
Often	1	**31**	**64**	4	109
Occasional	1	22	64	**13**	137
Never	0	25	75	0	4
Destination: abroad – percentage of answers					
Work-related trips	+increase	–diminish	0 no effect	Cannot say	n
Daily	5	**19**	71	5	42
Often	1	**21**	71	7	90
Occasional	2	13	68	**18**	111
Never	0	25	75	0	4

It seems that those Finnish office workers who make the majority of the work-related trips have experienced the diminishing potential of telematics most often when making trips in the metropolitan area, and less so the further they travel. Accordingly, the number of 'no effect' responses increases, especially regarding trips abroad.

Those who travel occasionally agree with those travelling daily or often that there is no effect of telematics. Those who travel occasionally are the majority of those who cannot say if telematics has had any effect. This finding seems to reduce the number of those having experienced a diminishing effect of telematics.

Reducing potential in non-routine trips

The personnel of companies make various types of trips and the need to travel for business depends on the branch of the company. The following trips were characterized as routine: visits within the company; visits to familiar clients and suppliers. Non-routine-type trips were excursions, trips to create new contacts, fair and conference trips (see Table 4.4). The routine trips are assumed to be more easily replaced by telematic means (see e.g. Törnqvist, 1993).

Table 4.4 Share of the most common work-related trips

Type of trip	%	(Törnqvist, 1993) %
Routine (**total**):	**49**	**64**
Visits within the company	18	24.1
Visits to clients	19	} 40.0
Making service calls or equipment installations	3	
Visiting municipal or governmental offices	9	
Non-routine (**total**):	**40**	**36**
Making new contacts	8	} 22.8
Excursions	9	
Having a stand at a fair, making a speech and consulting	5	
Fair visits, attending conferences and courses	18	13.2
Other	11	0

Although the number of trips for service calls or for equipment installations are few among the surveyed office workers (Table 4.4), they are most often mentioned as those that could be replaced by telematic services (Table 4.5). This is the case with the trips made in the metropolitan area and to other parts of the country.

It is interesting to discover that the well-educated office workers – half of them having an academic degree – claim that there is no necessity for differentiating between the purpose of the trip when considering the possibilities of using telematic services instead of making the trip. There seems to be in most cases (Table 4.5) no difference between trip type, whether it is made for routine or non-routine purposes, such as making new contacts and having a fair stand, making a speech or consulting activities. On the contrary, the expectation of the ease of replacing routine trips is not valid, especially when travelling abroad. After summarizing the diminishing effect of telematic services on the need of travelling according to the purpose of the trip, the

Table 4.5 Trip types where the diminishing effect of telecommunications on the work-related trips has been experienced

Type of trip	In the city area (%)	Other parts of the country (%)	Abroad (%)
Routine (**total**):	**44**	**48**	**43**
Visits within the company	12	13	12
Visits to clients	9	12	16
Making service calls or equipment installations	16	19	11
Visiting municipal or governmental offices	7	4	4
Non-routine (**total**):	**46**	**44**	**53**
Making new contacts	13	13	18
Excursions	9	6	4
Fair visits, attending conferences and courses	10	9	10
Having a stand at a fair, making a speech and consulting	14	16	21
Other	9	8	4

non-routine trips heading abroad are considered to have been replaced more often than routine trips abroad.

Transport mode

Work-related trips have been mainly made by car or public transport in the Helsinki Metropolitan Area. Aviation has been the major mode for the trips to other parts of Finland or abroad (see Table 4.6). Those who use public transport have experienced a less diminishing effect of telecommunications on their travelling than those using a car or aviation.

Means of information transfer

The office workers who had experienced a diminishing effect on travelling were asked which were the three telematic options most important for that trip reduction. First of all, regardless of the destination of the work-related trip – the one made in the metropolitan area or the one heading for a far-off destination – fax equipment has the leading position today in the reduction of work-related trips. E-mail and mobile phones seem to be natural followers of faxing as a means of replacing work-related trips (Table 4.7 and also Lehto, 1998a). Furthermore, the portable personal computer (PC) seems to have no

Table 4.6 Office workers' experience of the effect of telecommunications on work-related trips to different destinations according to travel mode

Destination: within the metropolitan area – percentage of answers

Mode	+increase	–diminish	0 no effect	Cannot say	n
Car	2	36	57	5	187
Public transport	0	31	62	7	95
Aviation	–	–	–	–	–
Other	20	0	65	15	20

Destination: other parts of the country – percentage of answers

Mode	+increase	–diminish	0 no effect	Cannot say	n
Car	1	35	56	8	98
Public transport	2	12	73	12	41
Aviation	1	30	62	7	118
Other	0	25	75	0	4

Destination: abroad – percentage of answers

Mode	+increase	–diminish	0 no effect	Cannot say	n
Car	25	0	50	25	4
Public transport	0	0	100	0	3
Aviation	2	22	69	7	173
Other	0	0	40	60	5

unconditional role in the reduction of the work-related trips despite its advantage in making the work independent of place. One use of portable PCs might be to free the personnel to work outside the office.

For comparison reasons, it is good to know that the highest impact on work effectiveness came from the everyday office tools: PC, phone, printer, fax and e-mail. Office workers favour personal printers and fax machines, but these services can be shared as well. Least important for work effectiveness are access and work-time control (Lehto, 1998a).

Telecoping and electronic mail affect all types of trips, while mobile phones are effective in replacing trips in the city area and when travelling to other

Table 4.7 Share of the device diminishing work-related trips, according to building type (IB=Intelligent building), gender and work position

Device	In the city area		In other parts of the country		Abroad	
Telefax	72		73		71	
E-mail	59		50		52	
Mobile phone	35		42		24	
Portable PC	16		15		13	
Answering machine	17		8		9	
Conference calls	5		7		5	
Videoconference	0.6		0.8		0	
Videophone calls	0.6		0		0	
Building type	IB	Non-IB	IB	Non-IB	IB	Non-IB
Telefax	71	74	72	73	**74**	64
E-mail	56	**62**	49	52	**57**	41
Mobile phone	**38**	33	39	**45**	25	23
Portable PC	**20**	13	17	13	9	**23**
Answering machine	17	17	**14**	0	9	9
Conference calls	2	**8**	6	8	**8**	0
Videoconference	0	1	0	2	–	–
Videophone calls	1	0	–	–	–	–

Gender	Male	Female	Male	Female	Male	Female
Telefax	70	75	71	78	75	60
E-mail	55	**64**	50	49	50	47
Mobile phone	**41**	27	43	43	25	20
Portable PC	**20**	11	**17**	11	13	13
Answering machine	11	25	7	8	8	7
Conference calls	2	9	6	5	4	13
Videoconference	0	2	0	3	–	–
Videophone calls	1	0	–	–	–	–

Working position	Supervisor	Expert	Clerk	Supervisor	Expert	Clerk	Supervisor	Expert	Clerk
Telefax	84	68	78	78	73	75	75	70	60
E-mail	63	55	59	50	49	50	50	44	60
Mobile phone	**53**	32	30	**50**	39	45	25	26	20
Portable PC	11	**20**	7	6	**18**	10	8	14	20
Answering machine	21	15	15	11	7	10	8	7	20
Conference calls	5	5	7	11	6	5	17	5	0
Videoconference	0	1	0	0	1	0			
Videophone calls	5	0	0	–	–	–			

parts of the country. Mobile phones are considered to have been somewhat less effective in replacing trips abroad (Table 4.7).

The differences between genders seem to reflect somewhat traditional roles, even though the data are not quite unambiguous (see Table 4.7). Men prefer mobility and rely upon mobile phones and portable PCs, while women rely upon devices used locally like e-mail and conference calls. This traditional thinking in the usage of telematics has also been apparent in studies on the use of telematics at home (Lehto, 1998b, 1999b).

Office workers' experience of telematics' influence on work-related trips was also compared in two different office building types: intelligent (IB) and non-intelligent (see also Lehto, 1996 and 1999a). In the IBs, workers more often report the diminishing effect, and in the non-intelligent offices no effect regarding trips to the other parts of the country or abroad (see Table 4.2). The telematically better working environment in IBs may have influenced the trips to more far-off destinations, but on the trips made in the metropolitan area this does not seem to be the case. No reason for this difference could be found from the data.

The potential of mobile phones and e-mail to reduce the number of work-related trips has also been proved by a study among municipal caretakers in central Finland (Knaapi and Luokola, 1993). The mobile phones save kilometres when caretakers know in advance if it is necessary to shop or to do other things on their way, instead of hearing of these tasks after coming to the client's place. Similar situations are familiar in other portions of the service sector as well. Computer printouts or on-site video can help maintenance people to take with them the necessary equipment without needing to check out the situation when they arrive on site.

Stimulation

The survey does not support expectations that telematics will generate new trips. Altogether, there are not many that report the experience of the increasing effect of telematics on work-related trips. None the less it is important to notice these kinds of signals and try to distinguish their message, especially in this case, where the hypothesis can be drawn from a related sector – the use of computers in office buildings. Instead of the forecast paperless office, paper consumption increased due to the increased use of computers.

Very few (1–2 per cent) of respondents think telematics has increased their travelling (see Table 4.2). These answers come from employees of companies that work with ICT or teletechnology. Due to the growth of their businesses, they have to travel to see customers and install equipment. This hypothesis can be further verified when noticing that these people are mostly heavy travellers (see Table 4.3) and they also often use travel mode 'other', which may refer to service vehicles (see Table 4.6).

One possible reason for this reported stimulation could theoretically lie in the use of mobile phones and portable PCs. These devices can transform the time used for travelling into effective working time with portable PCs or for communication time with the aid of mobile phones. This transformation reduces the cost of travel time.

Complementarity

It is interesting that those who travel most are male or female supervisors working in the intelligent offices (Lehto, 1998a). These professionals might have tried to find a working environment suitable for their tasks which require many (foreign) contacts and a suite of telemetric services. Supervisors prefer intelligent office features more than experts, and experts prefer them more than clerks (Lehto, 1996, 1999a and Lehto et al., 1997). Female supervisors in particular combine these two trends: heavy travelling and high appreciation of telematic devices. According to the study, the existence of the complementarity between telematics and travelling is obvious because the same workers, that is supervisors, travel a lot and take advantage of telematics as well.

The use of telematics is growing rapidly – much faster than travelling. In spite of obvious complementarity there probably exists invisible substitution. We do not know how big the growth of travelling would be without the telematically aided information exchange.

5. TELEMATICS AND SUSTAINABLE TRANSPORT POLICY

Transport policies are constrained by the dual demands for good mobility and environmental concern (see e.g. Dugonjik et al., 1993). The last 30 years have seen a general agreement that the amount of vehicle traffic ought to be restricted. However, in practice restrictions have been quite modest. In this chapter, we are considering work-related person trips which are mainly (measured in kilometres) made by cars and airplanes. While there is a long tradition of discussions about restricting car traffic, there has been less discussion of possible aviation restrictions.

A break point in the development of national information societies was US president Bill Clinton's proposal for the Information Superhighways programme in 1994. Many countries have followed with their own programme, encouraging ICT education, teleworking, industrial R&D work, social care

ICT solutions, research, and the like. Access to telenetworks has been seen as a democratic right for all citizens. Discussions on various points of view on this topic have been included; for example, in International Federation for Information Processing (IFIP) conferences and workshops all over the world, as well as on other occasions on ICT. Increasing use of telematic services has been regulated mainly through financial limitations and goal setting in companies; barring the standardization and *de facto* standardization in these realms, no discussions of restrictions – as in the transport sector – can be detected.

The focus in industrialization has been growth by means of technical and economic development. At the same time, people seeking such privileges have appealed to the idea of equal social rights and the limits of growth have been exceeded (see Figure 4.5.).

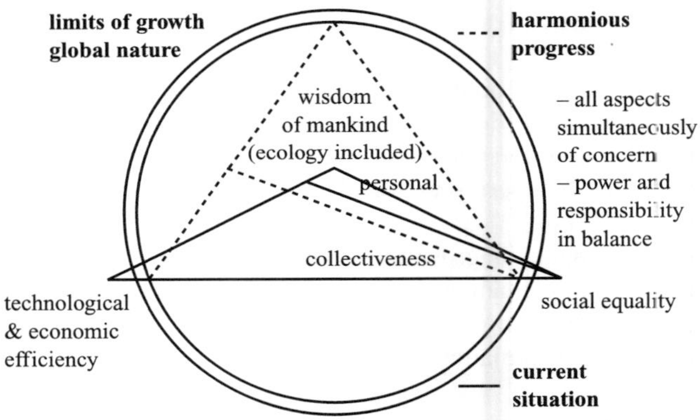

Figure 4.5 A model of holistic sustainability

It is obvious that, especially in work-related travelling and the use of telecommunications, money and time pose restrictions. The globalization of the world economy is increasing work-related travelling, especially over long distances. This shift is due to the increasing industrialization in the East, and the concomitant shift in the rest of the industrial world to an information society. Business transactions between information societies and the industrializing world still require meetings, even though much information can be transferred through ICT.

It is hard to imagine this type of industrial economy based solely on telex and telephone. In spite of the fact that transportation traffic is growing hand in hand with ICT, ICT must have had an invisible substituting effect on traffic already – otherwise this type of economy could not exist. Weak signals of this type of development have been seen among the office workers who answered the questionnaire. There are heavy travellers who also use a lot of ICT-aided working means.

In this context, two possible scenarios – based on a broad consideration of sustainability issues (see Figure 4.5) – can be formed:

(i) transport policy continues as before: much discussion about traffic restrictions, but very limited restrictions realized; or
(ii) severe restrictions on car traffic and aviation will be applied with the aid of various policy measures, for example pricing.

The first scenario does not affect current trends. The second scenario represents a revolution (see section 3) where the cost of travelling will be increased and the balance between travelling and telecommunications will consequently change. Then telecommunications can be used as a substitute for travelling and will have a clear role in the mitigation of the impacts of transport policy.

However, it must be remembered that the use of telecommunications (instead of travelling) is directed not only by the cost, but also by the impact which a message can have; for example, if you want to influence a person you must be able to note her reactions and realign your message accordingly. This is easier if you are in a face-to-face contact. Many other human behaviours influence the decision whether to take a trip or send an electronic message instead.

According to scientific tradition it is important to isolate factors from each other to prove the influence of separate factors. In a complex world this is not always possible. The concept of sustainable development includes the ecology of the material world, as well as the protection of the natural world, and human welfare and rights. Sustainable development problem solving seems to rely on multivariate solutions.

The authors of this chapter are highlighting the importance of human behaviour and attitudes in achieving harmonious and beneficial results in the interplay between travelling and the use of ICT. Without considering alternative approaches to the economy or balancing human activities in a new manner, maintaining the sustainable industrial economy is dependent on social equality and technological effectiveness. Sustainable development, that is, keeping within the limits of nature (see Figure 4.5), depends on human decisions, which are prudent to differing extents depending on the targets, the type of the problem, decision makers and decision-making procedure and so on. Thus, either social or technical development needs to be able to reach the harmony beneficial to both humans and nature.

In the first place, wisdom is cumulative across individuals. The wisdom of the society and policy making or legislation, rules and values can finally be traced to the wisdom of the individual. Trusting the end-user opinion, not only expert knowledge in research and technical development (RTD), but also in the dissemination of scientific and expert knowledge to all, is an important example of the means of aggregating wisdom – both individually and collectively – for the beneficial progress of the economy (and the information economy) to continue while keeping it, at the same time, within the limits of growth.

6. CONCLUSIONS

A **substitution** impact of telematics on work-related travelling can be seen in Finland. According to our survey, one-third of office workers (who regularly make work-related trips) had noticed it.

Very few, only 1 to 2 per cent, of the office workers had noticed any **stimulation** in work-related travelling due to the use of telematics.

About two-thirds of the office workers (who regularly make work-related trips) reported no impact of telematics on work-related travelling. Because they are also using telematics in their work, it can be stated that the majority considered that there exists a **complementarity** between telematics and travelling.

The office workers who did not regularly make work-related trips were more reluctant to state if telematics had any impact on travelling.

Because telematics has demonstrated an impact on travelling, it has that potential in the future. Thus, telematics can be used as a tool for the **mitigation** of the negative impacts of restrictive transport policy.

NOTE

1. Work-related trips are journeys for meetings, visiting customers and so on. They exclude trips between home and work.

REFERENCES

Ayres, R.U. (1990), 'Technological Transformations and Long Waves: Part II', *Technological Forecasting and Social Change*, **36**, 111–37.

Bruckmann, G. (1987), 'Will There be a Fifth Kondratief?', in T. Vasko (ed.), *The Long-wave Debate*, New York: Springer-Verlag, pp. 3–4.

Button, K. and R. Maggi (1995), 'Videoconferencing and its Implications for Transport: An Anglo-Swiss Perspective', *Transport Reviews*, **15**(1), 59–75.

Dugonjic, V., V. Himanen, P. Nijkamp and J. Padjen (1993), 'European Transport Policy: The Links between Mobility and Sustainability', in D. Bannister and J.

Berechman (eds), *Transport in a Unified Europe: Policies and Challenges*, Amsterdam: Elsevier, pp. 195–219.
Hienonen, R. and A. Lehtinen (1996), 'Elektroniikka- ja saehkoealan kehitysnaekymaet 1995–2000', *VTT Automaatio 1995*, Espoo (in Finnish).
Himanen, V., Kasanen, P. and Lehto, M. (1996), 'Information Transfer Saves Time and Trouble: Reduction of the Use of Energy for Work-related Passenger Transportation', LINKKI Publication 10/1996. Helsinki University (in Finnish; summary in English).
Himanen, V., Jaervi-Nykaenen, T. and Raitio, J. (1998), 'Daily Travelling viewed by Self-organizing Maps', in V. Himanen, P. Nijkamp and A. Reggiani (eds), *Neural Network Applications in Transport*, Aldershot: Avebury.
Hintikka, K.A. (1998), 'Puheenvuorojen kirjasto. Keskustelua suomalaisesta tietoyhteiskunnasta. SITRA 163', Helsinki.
King, A. and Schneider, B. (1991), *The First Global Revolution. A Report by the Council of the Club of Rome*, London, Simon & Schuster.
Knaapi, A. and Luokola, T. (1993), 'Kauha, kaennykkae, kodinhoitaja ... Tietoliikennepalveluja sosiaalitoimeen. Stakes', Raportteja 103, Helsinki.
Lehto, M. (1996), 'The "Intelligent Office" Concept Makes the Difference', *Nordic Architectural Studies Review. Nordisk Arkitekturforskning*, **1**, 61–72.
Lehto, M. (1997), 'The Resources of the Productive Office – Women's and Men's Opinions on Telematics and Intelligent Building Features', IFIP–WWC 97, Women, Work and Computerization, 'Spinning a Web from Past to Future'. Proceedings of the 6th International IFIP Conference, Bonn, Germany, 24–27 May, pp. 277–86.
Lehto, M. (1998a), 'Potential of Telematics in Profit Making, A Questionnaire in the Office Buildings in the Helsinki Metropolitan Area'. XV IFIP World Computer Congress '98. The Global Information Society on the Way to the NEXT Millennium. Telecooperation, The Global Office, Teleworking and Communication Tools. Vienna, Austria–Budapest Hungary. 31 Aug.–4 Sep. In R. Traunmüller and E. Csuháj-Varjú (eds), *Proceedings*, volume 121. Österreichische Computer Gesellschaft (Austrian Computer Society) on behalf of the International Federation for Information Processing (IFIP) CD-ROM edition.
Lehto, M. (1998b), 'Smart House Features in the Homes of Elderly – An End User Oriented Approach'. XV IFIP World Computer Congress '98. The Global Information Society on the Way to the NEXT Millennium. ICCHP '98. Sixth International Conference on Computers Helping People with Special Needs. Vienna, Austria–Budapest, Hungary. 31 Aug.- 4 Sep. In D.N.E. Alistair, E. Arato and W.L. Zagler (eds), *Computers and Assistive Technology: Proceedings of ICCHP '98*, volume 118. Österreichische Computer Gesellschaft (Austrian Computer Society) on behalf of the International Federation for Information Processing (IFIP), pp. 198–202 and CD-ROM edition.
Lehto, M. (1999a), 'Performance of the Intelligent Office Building in Profit Making. Intelligent and Responsive Buildings'. CIB Working Commission WO98. First International Congress. Brugge, Belgium, 29–30 March. In G. Klepfisch (ed.), *Proceedings*, Belgium Centre for Domotics and Immotics (BCDI), pp. 129–38.
Lehto, M. (1999b), 'Applicability of Smart House Features for Homes of Elderly. Intelligent and Responsive Buildings'. CIB Working Commission WO98. First International Congress. Brugge, Belgium, 29–30 March. In G. Klepfisch (ed.), *Proceedings*, Belgium Centre for Domotics and Immotics (BCDI), pp. 195–203.
Lehto, M., Flink, S., Karjalainen, S., Kovanen, K. and Pulakka, S. (1997), 'The Feasibility of the Intelligent Building Features in Office Buildings. The User

Oriented Quality, the Indoor Air and the Cost Effectiveness'. VTT Building Techn. Espoo. April (in Finnish).

Lehto, M. and Karjalainen, S. (1997), 'The Intelligent Building Concept as a User Oriented Approach to the Office – Focusing on Indoor Air and Energy Consumption'. Healthy Buildings/IAQ '97 Global Issues and Regional Solutions. Washington, DC, 28 Sep–2 Oct. In J.E. Woods, D.T. Grimsrud and N. Boschi, *Proceedings HB/ IAQ 97*, 3 volumes. Vol. 2, pp. 427–32.

Marchetti, C. (1987), 'On Transport in Europe: The Last 50 Years and the Next 20'. First forum on Future European Transport. Ludvig-Bölkow-Stiftung and Cesta, Munich.

Marvin, S. (1997), 'Environmental Flows. Telecommunications and the Dematerialisation of Cities?', *Futures*, **29**(1), 47–65.

Plaut, P.O. (1997), 'Transportation–Communications Relationships in Industry', *Transportation Research*, **A31** (6), 419–29.

Törnqvist, G. (1993), *Sverige I naetverkens Europa: Graensoeverskridandets former och villkor*. Malmoe: LiberHermods.

Vasko, T. (ed.) (1987), *The Long-wave Debate*, New York, Berlin, Heidelberg: Springer-Verlag.

APPENDIX 1 USE OF INFORMATION AND TELECOMMUNICATION TECHNOLOGY IN FINLAND – STATE OF THE ART

Finland is one of the small information-intensive nations among other Scandinavian countries, along with such countries as Ireland, Hungary, Israel, Hong Kong and Singapore. In these countries high-tech telematic solutions are used widely. There, the national telematic infrastructure is built in a reasonable time, while in the bigger ICT leading countries, the total system design and structuring is more time-consuming. The synergy obtained through the integration of various multimedia-aided telematic tasks accumulates faster in small societies. It is obvious that the final yield of the great nations will be vast. It remains to be seen if small nations will continue in the role of innovative competitors and specialists of spearhead technologies in the future.

Computing and telematics are used widely in various tasks in society. They are employed in business and at work, but are also used for leisure and entertainment, and in aiding caretaking. Use of telematics is essential in information-intensive work. Information is seen to be the fourth production force with capital, labour and machinery in the information age and the following service era.

Information rarely works alone, but is used with almost everything. In Finland, one in two of the workforce deals with information in their daily tasks to some degree. Flexi- or teleworking posts account for 9 per cent of all workers, and 10 per cent of those are full-time flexi-workers. People favour flexi-work to an increasing extent.

Finland and Sweden have the largest numbers of mobile phone users per capita in Europe (Hienonen and Lehtinen, 1996), and Finns access the internet more frequently than any other people in Europe. Of Finns, 50 per cent had a mobile phone in 1998 and every third household owned a home computer. Already by the end of 1996, one-third (33.8 per cent) of the Finns (between the ages of 15 and 74) were 'surfing' the internet (Hintikka, 1998).

Videoconferences (3 per cent once a year) and videophone calls (4 per cent once a year, 1 per cent weekly) are very seldom used. However, videoconferencing is more popular in office buildings provided with a videoconference room (Lehto et al., 1997 and Lehto, 1998). Video on desk is becoming inexpensive and more user-friendly. It might become more popular in the near future, following the practice of conference phone calls, which are made more frequently (25 per cent of respondents: 9 per cent daily, 3 per cent weekly, 5 per cent monthly and 8 per cent once a year).

Finnish R&D projects on telematic solutions in transport are many. They include travel and traffic information, telematics and sustainable development in an information society, variable message signs, traffic control and management, intelligent roads, cars, and parking, computer-aided navigation, human–machine interaction, safety and road-weather information systems, such as weather-controlled speed limits and displays and their effects on driver behaviour.

5. Telecommuting/teleworking: A virtual commuting possibility – the cases of Belgium and Brussels

Viviane Illegems, Alain Verbeke and Rosette S'Jegers

1. INTRODUCTION: THE RANGE OF CURRENT MOBILITY PROBLEMS IN BELGIUM AND BRUSSELS

The traffic congestion problem in Belgium and Brussels has increased substantially in the recent past. To measure the severity of the congestion problem, the growth in demand can be contrasted with the expansion of supply of road infrastructure. The demand for mobility can be analyzed using the following indicators: the number of motorized vehicles; the number of private cars; vehicle-kilometers and passenger-kilometers; and the vehicle occupancy rate. The length of the road network can be used as an estimate for the supply of road infrastructure. For the analysis of the tension between the demand for mobility and the supply of road infrastructure, the average traffic intensity per day[1] is used as a proxy. These indicators are described for 1985, 1990 and 1995 in Table 5.1 for Belgium and in Table 5.2 for Brussels. Every year, the number of registered motorized vehicles in Belgium increases substantially. Between 1985 and 1995, the number of registered motorized vehicles rose 28.5 percent. This rise is almost exclusively caused by an increase in the number of cars. In that same decade, the number of vehicle-kilometers increased 50 percent while the number of passenger-kilometers rose 39 percent. In that same period, a decrease of the average vehicle occupancy can be observed. The Belgian road network expanded 7.5 percent between 1985 and 1995. Expansion thus occurred at a lower pace than demand. This is reflected in the increase of the average traffic intensity per day for Belgium, by 39 percent between 1985 and 1995.

Traffic congestion in cities such as Brussels also increases continuously. Between 1985 and 1995, the number of vehicle-kilometers increased 22 percent and the passenger-kilometers 13 percent. In addition, the average vehicle occupancy rate in the Brussels Capital Region was characterized by a

Table 5.1 Mobility indicators for Belgium

Year	No. of motorized vehicles	No. of private cars	Vehicle-km per annum (bn)	Vehicle-km p.a. (bn) (1985 is reference)	Passenger-km p.a. (bn)	Passenger-km p.a. (bn) (1985 is reference)	Average vehicle occupancy	Length of road network in km	Average traffic intensity per day (6 a.m. – 10 p.m.)	Average traffic intensity per day (6 a.m. – 10 p.m.) (1985 is reference)
1985	3 840 792	3 342 704	53.64	100	70.79	100	1.54	133 153.5	1003	100
1990	4 454 884	3 864 159	70.28	131	89.54	126	1.49	139 076.0	1259	126
1995	4 936 084	4 273 451	80.26	150	98.50	139	1.44	143 175.0	1396	139

Sources: For number of private cars and number of motorized vehicles: Nationaal Instituut voor de Statistiek. Other data: Ministerie van Verkeer en Infrastructuur van Belgie.

Table 5.2 Mobility indicators for Brussels

Year	Vehicle-km p.a. (bn)	Vehicle-km p.a. (bn) (1985 is reference)	Passenger-km p.a. (bn)	Passenger-km p.a. (bn) (1985 is reference)	Average vehicle occupancy	Length of road network in km	Average traffic intensity per day (6 a.m. – 10 p.m.)	Average traffic intensity per day (6 a.m. – 10 p.m.) (1985 is reference)
1985	2.38	100	2.88	100	1.39	1621	3657	100
1990	2.73	115	3.18	111	1.34	1627	4179	114
1995	2.91	122	3.26	113	1.29	1643	4411	121

Source: Ministerie van Verkeer en Infrastructuur van Belgie.

further decline in that same period. Between 1985 and 1995 the length of the road network in Brussels did not increase significantly. The increased demand for mobility puts substantial pressure on the present road infrastructure. This situation is reflected in the 21 percent increase of Brussels average traffic intensity per day.

Every year more people in Brussels use private vehicles for commuting purposes. In 1981, 60.4 percent of the motorized moves took place with private vehicles; by 1991 this had increased to 72.9 percent. The increase of road congestion in urban areas such as Brussels is amplified by the spatial reorganization that took place in many Western European cities after World War II. This reorganization was based on the assumption that car use would increase. The migration from the city centers towards peripheral areas has also stimulated the use of private transport, because in these peripheral areas the public transport network has a lower density. Between 1981 and 1991 the number of jobs increased 2.2 percent in the Brussels Capital Region and 25.8 percent in the periphery. However, the number of inhabitants declined by 4.3 percent in Brussels and increased by 5.1 percent in the periphery (*Brussels Hoofdstedelijk Gewest*, 1995).

A reduction of the external effects that result from an increased demand for mobility can no longer be achieved through an increase in the physical infrastructure, since insufficient financial resources and/or space characterize most highly congested urban areas. As a result, governments are increasingly willing to adopt new policy instruments to reduce traffic congestion. A wide range of policies to reduce traffic congestion has been adopted in various countries. These vary from policies aimed at reducing traffic to policies that cope with stationary traffic. These policies may have an impact on local traffic flows as well as on regional and international traffic flows. Traditional traffic management instruments, such as restrictions on on- and off-street parking through physical, regulatory and pricing restrictions, traffic-calming measures and limiting access to roads for certain users, can clearly no longer deal with today's traffic congestion problems. Improved network performance via the application of new technologies will not solve the problem either since each of the existing systems (computer-linked traffic lights, direct vehicle-to-vehicle communications, the provision of up-to-date information and so on) has its limitations (Jones, 1993). Therefore, road pricing has often been proposed by transport economists as a response to congested roads. The principles of road pricing build upon Pigouvian tax principles. The road user should bear the real costs of the road transport externalities he causes. Such an optimal incentive structure should encourage the road user to change his/her behavior in all relevant aspects (for instance: number of moves; length of trips; technologies used; time of transfer and the like) (Verhoef, Nijkamp and Rietveld, 1996). However, despite the economic and welfare benefits that should result from this system, only a few successful applications of the concept exist, for example in Singapore, although at present several experiments are taking place in a large number of countries. The currently used anti-congestion policies undoubtedly have an important positive impact. The problem, however, is that a substantial impact sometimes only occurs when a combination of policies is pursued. Another problem is that such policies are only accepted by the public when linked to a program of expanded public transit, higher-quality transit, and road expansion. Unfortunately, the simultaneous pursuit of various anti-congestion policies requires substantial resources which government agencies often cannot afford to spend. In addition, most of these policies may take several years before they result in a reduction of congestion. Therefore, we suggest supplementing road pricing with a complementary policy, namely telecommuting/teleworking.

The question arises whether or not telecommunication is a substitute for transport. In this context several similarities between both can be observed:

1. *Physical similarity*: telecommunication networks are usually constructed next to existing road infrastructure.
2. *Analytical similarity*: for telecommunication as well as for transport, the presence of a network is an inherent requirement. The required networks also have a number of similarities, such as: links between segments with different capacities; and the possibility to use the network at different speeds and for different types of traffic.
3. *Economic similarity*: until recently, segments of telecommunication as well as segments of transport were considered natural monopolies. Only lately has this situation changed, largely as a result of technological developments.

Telecommunication has an influence on the demand for – and the supply of – transport facilities, but the reciprocal causal link also exists (Mokhtarian, 1990). The present chapter focuses on the impact of telecommunications in the sense of telecommuting/teleworking on the demand for transport. Telecommuting/teleworking as an anti-congestion policy tool can only be successful if a sufficient number of potential telecommuters/teleworkers is available. Individuals frequently describe their jobs as unsuitable for telecommuting/teleworking because a part of their job description requires their presence at headquarters. However, a wide range of tasks may exist that are highly suited to telecommuting/teleworking, for instance, administration, data processing, accountancy, writing of reports, systems analysis, consultancy, and so on. The general impression that telecommuting/teleworking is a full-time job no longer holds. One or more tasks in most jobs is suitable for telecommuting/teleworking. The fact that only a limited number of tasks included in most job descriptions are suitable for telecommuting/teleworking, combined with the new reality that this is true for most jobs, leads to the paradoxical conclusion that the potential to implement telecommuting/teleworking is much wider than usually thought. This can be called the 'Telecommuting/teleworking paradox'.

2. AN ALTERNATIVE WAY TO LOOK AT THE MOBILITY PROBLEM: TELECOMMUTING/TELEWORKING

The implementation of telecommuting/teleworking on a large scale could have a significant effect on road congestion, road accidents, environmental planning and the emission of pollutants. Telecommuting/teleworking is not a panacea for the current congestion and pollution problems. However, the promise of the reduction of road congestion could be realized in combination with complementary policies, such as road pricing, to cope with latent demand.[2] Telecommuting/teleworking as a strategy for reducing travel demand is attractive for several reasons:

1. If implemented, it has an immediate effect on demand.
2. It is relatively inexpensive to implement.
3. It increases personal choices rather than restricts them.
4. It has an impact on a variety of social, economic and ecological characteristics of society (Sampath et al., 1991):
   (a) social aspects: it offers the possibility to improve the integration of family life and work. The integration of disabled persons is also facilitated;
   (b) economic aspects: a well-designed telecommuting/teleworking project may result in a variety of benefits: increased productivity, a reduction in office space and overheads; improvement of recruiting facilities and the retaining of staff; reduction of absenteeism and higher quality customer services;
   (c) ecological aspects: as the penetration level of telecommuting/teleworking rises, the impact on traffic mobility, air quality and the development of rural areas increases.

Telecommuting/teleworking is especially interesting for urban areas like Brussels. Brussels has an ideal setting for the implementation of telecommuting/teleworking for a number of reasons. According to data of the Belgian Employment Department, 85 percent of the workforce is employed in the service sector. Jobs that are suited for telecommuting/teleworking are mostly situated in the service sector. Figure 5.1 reflects the distribution of telecommuters/

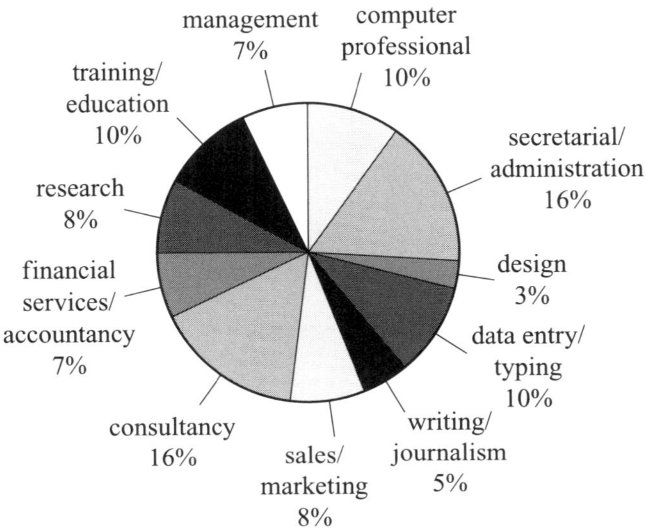

Source: Huws (1993).

Figure 5.1 *Distribution of telecommuters/teleworkers over the different professions in the UK*

teleworkers over the different professions in the UK. Commuting trips are responsible for 70 percent of the road congestion during peak hours (van Reisen, 1997). This results from the fact that during these periods, school, shop and work hours merge.

3. FORMS AND SCOPE OF TELECOMMUTING/ TELEWORKING

In order to assess the impact of telecommuting/teleworking on road congestion, the term must be clearly defined. In our research, we define telecommuting/ teleworking as the partial or total substitution of telecommunications and/or computer technology for the daily commuting to and from work (Handy and Mokhtarian, 1996). Telecommuting/teleworking can occur in the following forms:

1. *Organizational telecommuter*: the employee works on site, but is connected to other employees working on the same site or on the site of a supplier, customer or partner. This work is usually not regarded as

telecommuting/teleworking, because the commuting trip is not really eliminated.

2. *An employee who works at a satellite office*: the employee works within the physical boundaries of the company but is remote from central offices.
3. *An employee who works at a telework centre*: the employee works at a location situated in a residential area and owned by a third party that rents the facilities to various organizations.
4. *Electronic homeworker*: the employee works for a substantial part of the working hours at home on a regular basis, and is connected with his/her employer by telematics.
5. *Traditional homeworker*: the employee works for a substantial part of the working hours at home on a regular basis, but uses information technology only to transmit the results of his/her work.
6. *Nomadic worker*: the employee works at various sites according to current business needs.
7. *Professional networker*: the employee is a self-employed teleworker and delivers his/her services to more than one customer through the use of telecommunication modes (Huws, 1993).

Given the above classification of telecommuting/teleworking, the following measures can be assessed:

1. *Penetration*: the percentage of workers who telecommute/telework in the active workforce, without taking into account frequency.
2. *Levels of telecommuting/teleworking*: the percentage of workers who telecommute/telework on a particular day during the workweek (Handy and Mokhtarian, 1996).

The levels of telecommuting/teleworking can be expressed in three ways.

1. The number of workers who telecommute/telework as a percentage of the *potential* employment[3] on that particular day.
2. The number of workers who telecommute/telework as a percentage of the *actual* employment on that particular day. (The difference between (1) and (2) results from the fact that the actual employment does not include the individuals who are sick or on leave.)
3. The number of workers who telecommute/telework as a percentage of the *sum* of the number of commuting trips and the number of telecommuters/teleworkers. (Those individuals for whom home and workplace are the same are excluded from the analysis.) (Handy and Mokhtarian, 1996.)

4. CURRENT NUMBER OF TELECOMMUTERS/ TELEWORKERS IN BELGIUM AND BRUSSELS

At present no statistical data exist on the number of telecommuters/teleworkers in Belgium or Brussels. However, data concerning the number of homeworkers is available. These data simultaneously reflect an underestimation and an overestimation. They are an overestimation because they include individuals for whom the work and home environments are the same (for instance domestic workers and farmers). These individuals cannot be considered as telecommuters/teleworkers because no reduction of commuter trips took place. The data on homeworkers also lead to an underestimation because individuals employed in satellite offices or local telecommute/telework centers are not included (Handy and Mokhtarian, 1996).

According to data of the National Institute of Statistics, the total number of homeworkers in Belgium amounts to 628 029 individuals (Table 5.3) or 16.56 percent of the Belgian workforce. This number reflects both an underestimation and an overestimation for the reasons outlined above. A number of corrections was applied to obtain an approximation of the number of teleworkers/telecommuters in Belgium. The first correction also led to eliminating a number of sectors because the activities concerned are, in principle, not suited for telecommuting/teleworking or because the home and workplace are often identical. The sectors that were eliminated are NACE A/B (Agriculture, hunting and forestry/fishery), NACE F (Construction industry), NACE G (Wholesale and retail, car repair and household articles), NACE H (Hotels and restaurants), NACE K (Exploitation of and trade in real estate, renting and consultancy), NACE N (Public health and public assistance) and NACE P/Q (Private households with employees/Extraterritorial organizations and bodies).

Data on the exact number of homeworkers employed in NACE N and NACE K were not available in the sense that these data were collected in combination with data on the number of homeworkers in other sectors. For example, the homeworkers' data on sector N are included in a set that covers sectors L, M, N and O. It was then assumed that the proportion of homeworkers in sector N is the same as for the set of sectors L, M, N and O together. A similar calculation was done for NACE K. In the second step, the percentage of workers in the various activity classes ISCO 1 (Executive, administration, managerial), ISCO 2 (Professional specialists), ISCO 3 (Technicians and related sales) and ISCO 4 (Administrative support) was calculated *vis-à-vis* the total workforce in Belgium per sector. In most cases those four activity classes are well suited for telecommuting/teleworking. Here, the assumption was made that the ratio between ISCO 1, ISCO 2, ISCO 3 and ISCO 4 and the total Belgian employment is also found in the total number of homeworkers.

Table 5.3 Penetration level of telecommuting/teleworking in Belgium

	NACE C Extraction of minerals	NACE D Industry	NACE E Distribution of electricity gas and water	NACE I Transport, storage and communication	NACE J Financial institutions	NACE L/M/O Administration and defense, compulsory social security/education/ remaining welfare facilities and social-cultural and personal services	Total employment
Total number of individuals who usually work at home	259	21 558	218	3 457	17 120*	88 909*	382 556
Total number of individuals who sometimes work at home	386	25 435	1 592	6 310	16 024*	73 078*	245 473
Total number of homeworkers	645	46 993	1 810	9 767	33 144*	161 987*	628 029
Total number of individuals per sector according to occupation	9945	754 339	29 433	286 182	158 413	844 759	3 791 749

Total number of individuals in activity classes ISCO 1, ISCO 2, ISCO 3 and ISCO 4	4447	280 133	18 820	158 038	136 446*	589 746*	2 102 268
Total number of individuals employed in activity classes ISCO 1, ISCO 2, ISCO 3 and ISCO 4 as compared with the total workforce (in %)*	47.43	37.14	63.95	55.22	86.13	69.81	55.44
Total number of individuals self-employed	341	33 965	116	13 791	29 149*	65 645*	514 857
Total number of individuals self-employed as compared with the total workforce (in %)*	3.43	4.50	0.39	4.82	18.40	7.77	13.57
Total number of telecommuters/teleworkers*	295	16 668	1 112	5 133	23 294	104 297	150 799
Penetration level of telecommuters/teleworkers (in %)*	2.96	2.21	3.78	1.79	14.7	12.35	3.97

Note: * Own calculations; own compilation of table.

Source: NIS (1996).

The number of homeworkers was multiplied by this ratio. As a result, only ISCO 1, ISCO 2, ISCO 3 and ISCO 4 were maintained in the number of homeworkers. Finally the self-employed were removed from the total number of homeworkers. This correction is based on the following two assumptions:

1. The same percentage of self-employed individuals is found in all occupation classes.
2. The same ratio between the self-employed individuals and the total workforce is found in the total number of homeworkers.

For this correction, the ratio between the total number of self-employed individuals and the Belgian workforce was multiplied by the total number of homeworkers left after the previous correction. The result of the above calculations is an estimated penetration level of telecommuting/teleworking of 3.97 percent for Belgium. This is a very rough estimate which should only be used as an approximation. For Brussels, the same penetration level is assumed.

5. PROSPECTIVE NUMBER OF TELECOMMUTERS/ TELEWORKERS FOR BELGIUM AND BRUSSELS

To estimate the future number of telecommuters/teleworkers, the following approach can be used.

Trend Analysis

For this kind of analysis, the estimation of the current growth of the number of telecommuters/teleworkers is essential. This kind of data is unfortunately not available for Belgium or for the Brussels Capital Region. Still, this analysis can be applied because telecommuting/teleworking can be considered as a new commuting technology. According to the literature, the adoption of a new technology usually follows an S-curve. The shape of the S-curve is determined by the maximum penetration level and the growth of the number of telecommuters/teleworkers. The equation of the S-curve has the following form (Blackman, 1974):

$$\ln[P/(M-P)] = c_1 + c_2 \times t$$

with:

P = the current penetration level

M = the maximum penetration level

t = current year, where in the initial year $t = 0$
c_1, c_2 = constants.

Based on the assumption that 50 percent of the workforce are 'information workers' and that 80 percent of those workers are potential telecommuters/teleworkers, Nilles concludes that the maximum penetration of telecommuting/teleworking is 40 percent (Nilles, 1988). If this assumption is used for the Brussels and Belgian workforce, it appears that the activity classes ISCO 1 (Executive, administration, managerial), ISCO 2 (Professional specialists), ISCO 3 (Technicians and related sales) and ISCO 4 (Administrative support) represent 56.1 percent of the Belgian workforce and 72.9 percent of the Brussels workforce. Those four activity classes can roughly be viewed as including 'information workers.' In this context, the following comments can be made:

1. This group is bound to grow during the years to come.
2. Some of the workers identified as non-'information workers' may still be able to telecommute/telework part-time.
3. Not all 'information workers' will telecommute/telework, namely if their job is not well suited for it or if they do not consider it as an option.

To calculate the future penetration levels of telecommuting/teleworking in Belgium, the result of the preceding paragraph was used, namely a penetration of 3.97 percent for Belgium. Four scenarios were contemplated, based on four different growth percentages of telecommuting/teleworking, namely 2 percent, 5 percent, 10 percent and 20 percent (Tables 5.4 and 5.5). In the past decade, the number of telecommuters/teleworkers in the USA grew with 2.81 percent on a yearly basis (Handy and Mokhtarian, 1996). A similar, yet slightly reduced, growth rate was applied to the Belgian situation in the pessimistic growth scenario of 2 percent. The USA growth rate of 2.81 percent is possibly an underestimation, because only those telecommuters/

Table 5.4 Calculation of the parameter values for Belgium

1996–97	1996 adoption	$M-P$	1997 adoption	$M-P$	c_1	c_2	F
20%	0.0397	0.3603	0.0476	0.3524	–2.2056	0.2046	0.4
10%	0.0397	0.3603	0.0437	0.3563	–2.2056	0.1064	0.4
5%	0.0397	0.3603	0.0417	0.3583	–2.2056	0.0543	0.4
2%	0.0397	0.3603	0.0405	0.3595	–2.2056	0.0220	0.4

Source: Own calculations.

Table 5.5 Forecast of the penetration of telecommuting/teleworking in Belgium (all figures are percentages)

	Year									
Growth	1996	2001	2006	2011	2016	2021	2026	2031	2036	2041
20	3.97	9.38	18.41	28.14	34.73	37.93	39.23	39.72	39.90	39.96
10	3.97	6.32	9.68	14.08	19.22	24.47	29.13	32.81	35.44	37.19
5	3.97	5.05	6.38	7.97	9.85	12.00	14.39	16.98	19.67	22.37
2	3.97	4.38	4.83	5.32	5.84	6.42	7.03	7.69	8.40	9.15

Source: Own calculations.

teleworkers who telework at least three days per week were included. For the Belgian case a growth scenario of 5 percent per year was used. The current penetration level of telecommuting/teleworking in Belgium is lower than the USA penetration level. A catching up by Belgium therefore seems plausible. In addition, the growth rate of 5 percent was doubled twice, which resulted in growth scenarios of 10 and 20 percent on a yearly basis. The calculations were not made separately for Brussels and Belgium because for both geo-

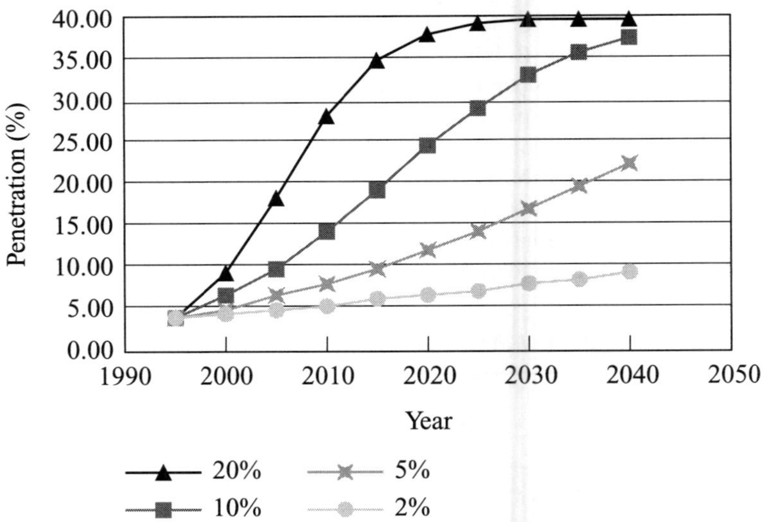

Source: Own calculations.

Figure 5.2 Evolution of the penetration of telecommuting/teleworking according to different growth scenarios for Belgium

graphic areas the same penetration level was assumed, leading to an identical outcome in each case.

From Table 5.5 the conclusion can be drawn that starting with a current penetration level of 3.97 percent and a growth of 20 percent, the maximum penetration will be reached in 2021, that is, after 25 years. If a growth rate of 10 percent occurs, approximately the same penetration level will be reached in 2041, after 45 years. The lower growth rates, namely 2 percent and 5 percent lead to that maximum penetration level, respectively after 200 years and 85 years. See also Figure 5.2.

If more statistical data were available concerning the growth of the penetration levels of telecommuting/teleworking, it would be possible to determine which of the scenarios above is the most realistic one. It is clear that policies promoting the implementation of telecommuting/teleworking may influence the likelihood that specific scenarios will take place in practice.

Analysis of the Characteristics of a Telecommuter/Teleworker

A forecast of the trend in the number of telecommuters/teleworkers can also be based on the characteristics of those workers who are highly correlated with the decision to telecommute/telework. For instance, when it is taken into account that persons in their thirties have a high willingness to telecommute/telework and that this group is bound to grow during the next decade, this leads to the conclusion, *ceteris paribus*, that the number of telecommuters/teleworkers will grow. Given that the decision to telecommute/telework is related to professional occupation, this makes the 'profession' a promising characteristic. In general terms, the following occupations are considered suitable for telecommuting/teleworking: Executive, administration, managerial; Professional specialists; Technicians and related sales and administrative support. The shift–share technique can be used to analyze the anticipated effect on the number of telecommuters/teleworkers resulting from a change in the distribution of the workers over the different occupation classes (Handy and Mokhtarian, 1996). Originally, this technique was used to analyze the change in employment in a region as compared to the national employment for every branch of industry. This technique provides insights into the growth of telecommuting/teleworking penetration caused by a change in the distribution of the workforce over the different occupation classes. The potential growth of the number of telecommuters/teleworkers through a change in the distribution among the different occupation classes can be divided into a national, a regional and an occupation effect. The national effect reflects the global growth of the national employment. The occupation effect gives the growth in an occupation class after a deduction of the national effect. The regional effect reflects the growth in employment in a region for an occupa-

PRIFYSGOL CAERDYDD

tion class after a deduction of the national growth calculated for that particular occupation class. The occupation classes used by the National Institute of Statistics of Belgium were divided into occupation classes suitable and not suitable for telecommuting/teleworking. For this analysis, data from NIS (1993) and (1996) were used. These data were used because the classification of the occupations is based on international standards. Table 5.6 reflects the growth of the different occupation classes between 1993 and 1996 for the

Table 5.6 Trends in occupations in the Brussels Capital Region

	1993		1996		Growth (%)
	Total	%	Total	%	
Executive, administration, managerial	76 407	12.15	85 404	13.63	11.78
Professional specialists	130 429	20.74	140 579	22.43	7.78
Technicians and related sales	72 039	11.46	61 820	9.87	–14.19
Administrative support	165 123	26.26	169 238	27.01	2.49
Sub-total	**443 998**	**70.61**	**457 041**	**72.94**	**2.94**
Service occupations	52 433	8.34	52 037	8.30	–0.76
Agriculture, forestry and fishing and related	2 152	0.34	932	0.15	–56.69
Others	130 185	20.71	116 602	18.62	–10.43
Sub-total	**184 770**	**29.39**	**169 571**	**27.06**	**–8.226**
Total	**628 768**	**100**	**626 612**	**100**	**–0.34**

Table 5.7 Trends in occupations in Belgium

	1993		1996		Growth (%)
	Total	%	Total	%	
Executive, administration, managerial	372 757	10.0	408 021	10.90	9.5
Professional specialists	665 135	17.8	709 030	18.93	6.6
Technicians and related sales	355 551	9.5	353 762	9.45	–0.5
Administrative support	631 123	16.9	631 455	16.86	0.1
Sub-total	**2 024 566**	**54.1**	**2 102 268**	**56.14**	**3.8**
Service occupations	359 152	9.6	380 029	10.15	5.8
Agriculture, forestry and fishing and related	103 389	2.8	105 214	2.81	1.8
Others	1 216 264	32.5	1 157 073	30.90	–4.9
Sub-total	**1 678 805**	**44.9**	**1 642 316**	**43.86**	**–2.2**
Total	**3 741 702**	**100**	**3 785 905**	**100**	**1.2**

Table 5.8 Shift–share analysis of the growth in occupations suited for telecommuting/teleworking

	Growth (%)			Shift–share analysis (%)		
	(1)	(2)	(3)	(4)	(2) – (1)	(3) – (2)
Executive, administration, managerial	1.2	9.5	11.78	1.2	8.3	2.28
Professional specialists	1.2	6.6	7.78	1.2	5.4	1.18
Technicians and related sales	1.2	–0.5	–14.19	1.2	–1.7	–13.69
Administrative support	1.2	0.1	2.49	1.2	–1.1	2.39

Notes: (1) = growth in Belgian workforce; (2) = growth in national occupation classes; (3) = growth in employment divided over the occupation classes for the Brussels Capital Region; (4) = (1) = national effect; (2) – (1) = effect following from the growth in occupation classes; (3) – (2) = effect following from the Brussels Capital Region.

Source: Own calculations.

Brussels Capital Region and Table 5.7 gives the same information for Belgium. Table 5.8 displays the results of the shift–share analysis.

Table 5.8 should be interpreted as follows: the increase of 11.78 percent for the occupation class Executive, administration, managerial in the Brussels Capital Region comprises the 1.2 percent growth in national employment, the 8.3 percent growth in that particular occupation class and the 2.28 percent extra growth in employment in the Brussels Capital Region compared with national employment. The occupational effect is positive for two of the four occupation classes suited for telecommuting/teleworking, namely Executive, administration, managerial and Professional specialists. Only for the occupation class Technicians and related sales does the fact that a person is employed in the Brussels Capital Region have a negative influence on the employment in that occupation class.

The implications of the application of the growth in employment according to the occupation class for Brussels on the penetration level of the number of telecommuters/teleworkers in Brussels can be found in Table 5.9. The penetration level of telecommuting/teleworking calculated earlier on for Belgium, namely 3.97 percent, is also used as penetration level for Brussels. The employment in occupations suited for telecommuting/teleworking represents 72.94 percent of the active workforce. Hence it is assumed that: (a) 5.44 percent (3.97%/0.7294) of the persons employed in jobs suited for telecommuting/teleworking are telecommuters/teleworkers; whereas (b) this is 0 percent for the occupations not suited for telecommuting/teleworking; and (c) the same ratio is found in 2004, eight years later. The growth for each occupation class

Table 5.9 Forecast of the penetration level in the Brussels Capital Region based on the shift–share approach

	Growth of the number of telecommuters/teleworkers 1997–2004 (%)	Telecommuters/teleworkers in the occupation classes (%)	1996		2004	
			Number of individuals employed	Number of telecommuters/teleworkers	Number of individuals employed	Number of telecommuters/teleworkers
Professions suited for telecommuting/teleworking						
Executive, administration, managerial	23.56	5.44	85 404	4 646	105 525	5 741
Professional specialists	15.56	5.44	140 579	7 647	162 453	8 837
Technicians and related sales	–28.38	5.44	61 820	3 363	44 275	2 409
Administrative support	4.98	5.44	169 238	9 207	177 666	9 665
Professions not suited for telecommuting/teleworking	–16.45	0	169 571	0	141 677	0
Total workforce	–0.64		626 612	24 863	631 596	26 652
Penetration level of telecommuting/teleworking (%)				3.97		4.22

Source: Own calculations.

between 1993 and 1996 was doubled for the period 1997–2004 because the time considered is twice as long.

For a more exact forecast, data on the percentage of telecommuters/teleworkers per occupation class and data on the evolution of the number telecommuters/teleworkers per occupation class are a necessity. According to the shift–share analysis a quicker growth of the jobs suited for telecommuting/teleworking compared to the growth in jobs not suited for telecommuting/teleworking will lead to an increase of 0.25 percent of the number of telecommuters/teleworkers in Brussels between 1997 and 2004.

Analysis of the Decision to Telecommute/Telework

Starting from a model that reflects the individual decision to implement telecommuting/teleworking or not, the chance that an individual with certain characteristics in a certain situation will telecommute/telework can be assessed. The future number of telecommuters/teleworkers can be predicted based on the evolution of those characteristics and environmental factors that have a significant positive effect on the decision to telecommute/telework. Through this model of individual choice, the impact of policy tools on the number of telecommuters/teleworkers can be assessed.

Transportation Forecasting Models

The future number of telecommuters/teleworkers can also be predicted, based on the inclusion of the telecommuting/teleworking option in the transportation forecasting models. These models focus on trip distribution, mode split, route choice and trip distribution. The emphasis is placed on the trip, not on the activity. This is the reason that telecommuting/teleworking is only treated implicitly. The analysis is based on data about every trip during a period obtained from a travel diary survey. For telecommuters/teleworkers a lower number of trips is registered. When traffic is forecast for a region, the frequency of telecommuting/teleworking has a negative influence on the number of trips per employee. In these models the assumption is made that the average number of trips per occupation class per region remains constant. The fact that growth in penetration of telecommuting/teleworking could reduce the number of trips per occupation class even more in the future is not taken into account by transportation forecasting models when the traffic for a region is forecast. Another limitation of the transportation forecasting models is that the margin of error for these models lies between 10 and 20 percent. At the moment, the frequency of telecommuting/teleworking in Belgium is low. This makes it plausible that the net impact of telecommuting/teleworking on the number of trips will not be significant according to these models.

6. IMPACT OF TELECOMMUTING/TELEWORKING ON TRANSPORT

The main effect of telecommuting/teleworking on mobility is the reduction of commuter trips during peak hours. It will also influence non-work-related trips. The effect on non-work-related trips can be positive (for instance increased use of public transport or the bicycle) as well as negative (for instance the termination of car-pool initiatives) (Bernardino et al., 1992).

A method developed by Olszewski and Lam was used to gain insights into the daily reduction in trips in Belgium and Brussels (Olszewski and Lam, 1996). This method only considers work-related trips and assumes that the modal split for telecommuters/teleworkers is identical to the modal split for other employees. The method of Olszewski and Lam gives an estimation of the reduction in motorized commuter round trips and vehicle commuter round trips caused by a specific penetration level and frequency of telecommuting/teleworking. For the analysis the following steps must be taken:

1. Determination of the total number of motorized commuter round trips for Belgium and the Brussels Capital Region.
2. Determination of the total number of vehicle commuter round trips for Belgium and the Brussels Capital Region.
3. Determination of the reduction in the number of trips caused by the implementation of telecommuting/teleworking in Belgium and the Brussels Capital Region.

Determination of the Motorized Commuter Round Trips (*M*)

Motorized commuter round trips should be understood as every motorized trip, private as well as public: $M = B \times t$, where B = the workforce and t = the number of motorized commuter round trips per employee per day.

The coefficient t is lower than one because absences because of strikes, leave and sickness have to be taken into account. In addition employees who walk to work or who are self-employed, which means they mostly work where they live, should not be considered. The first step in the calculation of the exact t coefficient is the determination of the correction for strikes, leave and sickness. Starting from an average of 20 work days per month, a Belgian employee works 264 days a year. But, every Belgian employee is entitled to a least 22 vacation days and 11 public holidays. A Belgian employee thus has at least 31 days of leave a year. The same conclusion can be drawn for Brussels. Based on data of Deliege (1993), the average number of sick days for a Belgian employee can be calculated. On average, a Belgian employee is sick 7.7 days in one year. We assume that the health pattern for the Brussels employees is the same as for the Belgian employees. For the calculation of the number of strikes per employee data from the *Statistisch jaarboek van België* (1994) were used. A total of 153 706 days of strike were registered for the Belgian workforce, or 0.004 days for every employee. This percentage was also adopted for Brussels. This leads to the fact that, on average, a Belgian as well as a Brussels employee does not work 264 days a year but 225.26 days a year (264 – 31 – 7.7 – 0.004), and this results in a correction of t towards 0.8533 (the number of actual workdays as compared to the number of official working days). The removal of the self-employed is the second correction that is made. It appears that 13.57 percent of the Belgian workforce is self-employed and 86.42 percent is not self-employed. This results once more in a correction of t for Belgium towards $t = 0.7374$. In Brussels, 17.08 percent of the workforce is self-employed and 82.92 percent of employees are not self-employed. When the correction for sickness, leave and vacation is combined with the correction for self-employment for Brussels a t of 0.7076 is found (NIS, 1996). A total of 5.3 percent of the trips of the active Belgian workforce and 4.9 percent of the trips of the active Brussels workforce are made on foot (*Volks- en Woningtelling*, 1991). The last correction results in a coefficient t of 0.6983 for Belgium ($t_{\text{Belgium}} = 0.6983$) and a coefficient t of 0.6729 for Brussels ($t_{\text{Brussels}} = 0.6729$). The workforce amounts to 3 794 552 in Belgium ($B_{\text{Belgium}} = 3\ 794\ 552$) and 628 736 in Brussels ($B_{\text{Brussels}} = 628\ 736$). This results in a total number of motorized commuter round trips of 2 649 727 for Belgium ($M_{\text{Belgium}} = 2\ 649\ 727$) and 423 076 for Brussels ($M_{\text{Brussels}} = 423\ 076$).

Determination of the Number of Vehicle Commuter Round Trips (*V*)

The number of vehicle commuter round trips only includes the motorized commuter round trips that take place with a private vehicle and this number is also corrected for car-pooling.

$$V = (B \times t \times p) / O = (M \times p) / O$$

where p = the percentage of private traffic as compared to the total motorized trips and O = the average vehicle occupancy.

In Belgium private transport is used for 86.53 percent of the traffic ($p_{\text{Belgium}} = 0.8653$) (*Algemene Verkeerstellingen 1995,* 1997). This percentage is also used for Brussels ($p_{\text{Brussels}} = 0.8653$). The average vehicle occupancy in Belgium is 1.44 persons per vehicle ($O_{\text{Belgium}} = 1.44$), while for Brussels this is 1.29 persons per vehicle ($O_{\text{Brussels}} = 1.29$) (ibid.). Consequently the number of vehicle commuter round trips amounts to 1 592 228 in Belgium (V_{Belgium} = 1 592 228) and to 283 789 in Brussels (V_{Brussels} = 283 789).

Determination of the Reduction in Trips Caused by the Implementation of Telecommuting/teleworking

First the reduction in the number of motorized commuter round trips is assessed (M_r):

$$M_r = M \times a \times f \times m$$

Then the reduction in the number of vehicle commuter round trips is computed (V_r):

$$V_r = V \times a \times f \times m$$

where

a = the current penetration level as compared to the maximum penetration level,

f = the average frequency with which one telecommute/teleworks, in the context of a five-day workweek; and

m = maximum penetration level of telecommuting/teleworking

For the current penetration level in Belgium, data obtained from the application of the S-curve for 1996, 2011 and 2021 for a low growth rate of 5 percent and a high growth rate of 20 percent were used (see trend analysis). The same data for Belgium are used for Brussels. The data obtained through the trend analysis must be rewritten taking into account the maximum penetration level of 40 percent. At this moment no data exist on the frequency of telecommuting/teleworking in Belgium and the Brussels Capital Region. The frequency in the United States is estimated at 2 days per workweek (Handy and Mokhtarian, 1996) and Olszewski and Lam also estimate the frequency for Singapore at 2 days per workweek (Olszewski and Lam, 1996). For The Netherlands, the frequency has been estimated between 1 and 1.5 days per workweek (van Reisen, 1997). Because the Belgian situation is best compared with the Dutch situation, the average frequency for Belgium and Brussels is assumed to be 1.5 days per workweek. The maximum penetration level is 40 percent for the reasons mentioned earlier. The results of the calculation of the reduction of motorized commuter trips and vehicle commuter trips are presented in Table 5.10 for Belgium and in Table 5.11 for Brussels. The total number of motorized and vehicle commuter round trips is assumed to be constant for the three points of time.

Table 5.10 Reduction in trips caused by implementation of telecommuting/teleworking in Belgium

	Penetration level of telecommuting/teleworking rises yearly by 5%			Penetration level of telecommuting/teleworking rises yearly by 20%		
	1996	2011	2021	1996	2011	2021
Penetration level of telecommuting/teleworking with regard to the Brussels workforce (%)	3.97	7.97	12.00	3.97	28.14	37.93
Motorized round commuter trips (M)	2 649 727	2 649 727	2 649 727	2 649 727	2 649 727	2 649 727
Vehicle round commuter trips (V)	1 592 228	1 592 228	1 592 228	1 592 228	1 592 228	1 592 228
Reduction in the motorized round commuter trips (M_r)	31 558	63 355	95 390	31 558	223 690	301 512
Reduction in the vehicle round commuter trips (V_r)	18 963	38 070	57 320	18 963	134 416	181 180
Reduction in the motorized round commuter trips with regard to total number of motorized round commuter trips (%)	1.2	2.4	3.6	1.2	8.4	11.4
Reduction in the vehicle round commuter trips with regard to the total number of vehicle round commuter trips (%)	1.2	2.4	3.6	1.2	8.4	11.4

Source: Own calculations.

Table 5.11 Reduction in trips caused by implementation of telecommuting/teleworking in the Brussels Capital Region

	Penetration level of telecommuting/teleworking rises yearly by 5%			Penetration level of telecommuting/teleworking rises yearly by 20%		
	1996	2011	2021	1996	2011	2021
Penetration level of telecommuting/teleworking with regard to the Brussels workforce (%)	3.97	7.97	12.00	3.97	28.14	37.93
Motorized round commuter trips (*M*)	423 076	423 076	423 076	423 076	423 076	423 076
Vehicle round commuter trips (*V*)	283 789	283 789	283 789	283 789	283 789	283 789
Reduction in the motorized round commuter trips (M_r)	5 039	10 116	15 231	5 039	35 716	48 142
Reduction in the vehicle round commuter trips (V_r)	3 380	6 785	10 216	3 380	23 957	32 292
Reduction in the motorized round commuter trips with regard to total number of motorized round commuter trips (%)	1.2	2.4	3.6	1.2	8.4	11.4
Reduction in the vehicle round commuter trips with regard to the total number of vehicle round commuter trips (%)	1.2	2.4	3.6	1.2	8.4	11.4

Source: Own calculations.

From Table 5.10 it can be deduced that if telecommuting/teleworking reaches a significant level the impact on traffic congestion will be significant. According to the model used, a penetration level of telecommuting/teleworking of 37.93 percent would lead to a reduction of the number of commuter trips by 11.4 percent or 181 180 vehicle commuter round trips and a penetration level of 12 percent would result in a decrease of 3.6 percent in commuter trips or 57 320 vehicle commuter round trips.

A similar conclusion as for Belgium can be drawn for the Brussels Capital Region (see Table 5.11).

This rough approach to estimating the reduction of commuter trips in Belgium and Brussels confirms the potential of telecommuting/teleworking as a policy tool for the reduction of traffic congestion. The data used in this analysis are, however, very rudimentary.

7. CONCLUSIONS AND RESEARCH NEEDS

Given that traffic congestion problems in Belgium and Brussels are increasing, the potential of telecommuting/teleworking as an anti-congestion policy tool merits further investigation. At present no statistical data are available on the number of telecommuting/teleworking individuals in Belgium and the Brussels Capital Region. A rough estimation of the current number of telecommuting/teleworking individuals was obtained after performing several corrections on the data about the number of homeworkers in Belgium. As a result of these corrections a penetration level of 3.97 percent was determined for Belgium and Brussels. Extensive data collection is a prerequisite for a more accurate analysis of the impact of telecommuting/teleworking on traffic congestion and for a correct calculation of current penetration levels and frequency of telecommuting/teleworking. A survey of the methods used to assess the number of telecommuters/teleworkers was given, namely trend analysis, analysis of the characteristics of a telecommuter/teleworker, analysis of the decision to telecommute/telework and transportation forecasting models. The application of these methods to assess the number of telecommuters/teleworkers is made difficult by a lack of sufficient and/or accurate data. Finally, a rough estimation was made of the reduction in commuter trips that would result from the implementation of telecommuting/teleworking. An application of the Olszewski and Lam model suggests that a penetration level of telecommuting/teleworking of 12 percent leads to a reduction of the commuter trips by 3.6 percent. The potential of telecommuting/teleworking as an anti-congestion policy tool should be investigated using more accurate statistical data.

NOTES

1. The average traffic intensity per year is calculated by dividing the total vehicle-kilometers in one year in Belgium by the total length of the road network in that year. The average traffic intensity per year is then divided by 365, which results in the average traffic intensity per day. Finally the calculated value is divided by 1.1, which results in the average traffic intensity per day between 6 a.m. and 10 p.m. This is based on the assumption that the traffic volume between 10 p.m. and 6 a.m. is one-tenth of the volume between 6 a.m. and 10 p.m. If the traffic volume remained constant during 24 hours, then a coefficient of 1.5 would need to be used rather then a coefficient of 1.1. A similar reasoning is followed for the calculation of the traffic intensity per day between 6 a.m. and 10 p.m. for Brussels.
2. This concept refers to individuals who used public transport or other low-impact modes to travel in the past, but now use private vehicles due to a reduction in road congestion.
3. Potential employment includes all jobs, even those that are not filled because of sickness or vacation.

REFERENCES

Algemene Verkeerstellingen 1995 (1997), Ministerie van Verkeer en Infrastructuur, 11.

Bernardino A., M. Ben-Akiva and I. Salomon (1992), 'Stated Preference Approach to Modeling the Adoption of Telecommuting', *Transportation Research Record*, **1413**, 22–30.

Blackman, A. (1974), 'The Market Dynamics of Technological Substitution', *Technological Forecasting and Social Change*, **6**(1), 41–63.

Brussels Hoofdstedelijk Gewest: Gewestelijk Verkeerscirculatieplan (1995), STRATEC.

Deliege, D. (1993), *Les Absences pour Maladie dans le Secteur Privé*, Louvain-La-Neuve.

Handy, S. and P. Mokhtarian (1996), 'Forecasting Telecommuting', *Transportation*, **23**, 163–90.

Huws, U. (1993), 'Teleworking in Britain: A Report to the Employment Department', *Research Series*, **18**, 70.

Jones, P. (1993), 'Congestion and the Car, Transport in Unified Europe Policies and Challenges', *Studies in Regional Science and Urban Economics*, **24**, 29–53.

Mokhtarian, P. (1990), 'A Typology of Relationships between Telecommunications and Transportation', *Transportation Research A*, **24**(3), 231–42.

Nilles, J. (1988), 'Traffic Reduction by Telecommuting: A Status Review and Selected Bibliography', *Transportation Research A*, **22A**(4), 301–17.

NIS (1996), *Enquête naar de beroepsbevolking*, Sociale Statistieken, Nationaal Instituut voor de Statistiek, Ministerie van Economische Zaken.

NIS (1993), *Enquête naar de beroepsbevolking*, Sociale Statistieken, Nationaal Instituut voor de Statistiek, Ministerie van Economische Zaken.

Olszewski, P. and S.H. Lam (1996), 'Assessment of Potential Effect on Travel of Telecommuting in Singapore', *Transportation Research Record*, **1552**, 154–60.

Sampath, S., S. Saxena and P. Mokhtarian (1991), 'The Effectiveness of Telecommuting as a Transportation Control Measure', *Proceedings of the ASCE Urban Transportation Planning and Air Quality*, **44**, 347–62.

van Reisen (1997), *Ruim baan door telewerken? Effecten van flexibele werkvormen*

op ruimtelijke ordening en mobiliteit als gevolg van veranderend tijd-ruimtegedrag, Delft: Publikatiebureau Bouwkunde, 229.
Verhoef, E., P. Nijkamp and P. Rietveld (1996), 'Tradeable Permits: Their Potential in the Regulation of Road Transport Externalities', Third NECTAR Euroconference, Mons.
Statistisch Jaarboek van België (1994), 112, Nationaal Instituut voor de Statistiek, Ministerie van Economische Zaken.
Statistisch Jaarboek (1994), Belgacom.
Volks- en Woningtelling (1991), Nationaal Instituut voor de Statistiek.

6. Long-distance telephone calls, media endowment and contact network: An empirical study

Rico Maggi and Alessandro Cento

1. THE PROBLEM

The evidence presented in this chapter stems from a broader research project into the determinants of the adoption and use of new media. More specifically, we analyse the influence of economic factors, individual attitudes and contact networks. In order to understand individual behaviour with respect to new media, it seems useful to analyse behaviour with respect to existing media first, taking the same frame of reference. As a consequence, we concentrate in a first step on the analysis of telephone use. The endowment (represented by equipment) enters only in the form of exogenous variables representing an individual's past decisions.

While the adoption and use of new media is often treated as something totally new, here it is proposed to conceive of it as a new behaviour with respect to the same thing (that is, telecommunication). Telecommunication as an activity is hence at the centre of interest of this study. In the vein of the literature on household production (see Muellbauer, 1974; Epple, 1987), telecommunication can be 'produced' by the individual in different ways in order to maximize utility. Telecommunication here includes all forms (one-way, two-way, any medium) of non-mass communication for any purpose (tele-shopping, tele-commuting, simple phone conversations, and so on). The advantage with this perspective is that demand for new media depends on the way an individual intends to use them for 'producing telecommunication'. An individual can produce telecommunication with different technologies by choosing a set of media for communicating with members of her contact network. The notion of contact network is a second critical concept of this research. By contact network we mean a set of one-to-many links established by an individual by way of investments (of time, money and effort) into a permanent relation. This decision is not modelled here; rather, the contact network is given and treated as exogenous in the model, but it is believed to be of relevance for telecommunication demand.

New technologies change the production possibilities for telecommunication. Depending on the purpose, the contact network, individual preferences, characteristics, income and attitudes, an individual will decide on the adoption and use of different media by considering their price and quality (factor demand). Hence, starting from the traditional telephone mode, we want to understand the way in which the availability of different production possibilities among individuals influences their telecommunication behaviour in the form of phone calls. Different production possibilities in this context are represented by the potential use of alternative media (cellular phones, fax, e-mail, and so on).

The simple model of household production serves to derive demand functions for phone calls, which are then tested and discussed. The chapter thus allows for a first test of the analytical framework proposed here.

2. THE MODEL

The model we use at present is restricted to telephone use, given a set of media, a personal contact network, a vector of individual characteristics and a set of media prices.

The decision context can be presented as in Figure 6.1, where the upper part contains the decisions on contact network and media adoption, which are not considered in this chapter. The same holds for the feedback from telecommunication demand on contacts and endowment. The part to be modelled is hence the one indicated by the solid-line arrows, that is, the dependence of telecommunication demand on individual characteristics and attitudes, size and structure of the personal contact network, the endowment with other media and the supply of services. In the present context of phone calls, the latter are described simply by the prices for peak and off-peak time, but the concept will later be extended to include media characteristics.

We model the decision behaviour of an individual in the short run (endowment and contact network are fixed and exogenous). In accordance with the theory of household production, the decision is conceived as a two-stage optimization procedure (see, for example Filippini, 1995) where an individual maximizes utility from telecommunication which first has to be produced in an optimal way. In this formulation individuals are both producing units and utility maximizers.

Thus, we assume that an individual first resolves a cost minimization problem which, given the above context, takes the following form:

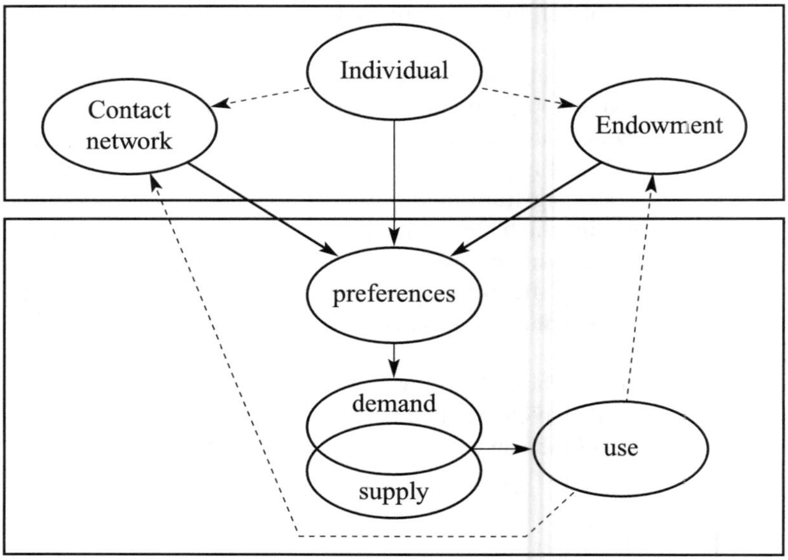

Figure 6.1 Decision context for telecommunication

$$C_t = c(T, P_d, P_n, \tau_T, R, E_d, E_{id}) \tag{6.1}$$

The individual minimizes the cost C_t of producing a certain quantity of telecommunications T in the short run, given the prices P_n P_d , of the variable factors (that is, the number of peak and off-peak calls), τ_T as the unit of time input used for producing the commodity T and the quantities of the fixed factors (contact network R, endowment with other media E_d, E_{id}). We assume that individual has already decided to allocate her total time working and consuming, and then this decision does not enter in the model. Time spent consuming is allocated for telecommunication and other goods consumption.

$$\bar{\tau}_C = \bar{\tau}_T + \bar{\tau}_X \tag{6.2}$$

where $\bar{\tau}_T$ is the telecommunication time, $\bar{\tau}_C$ is the other goods consumption time.

As the notation (6.1) indicates we distinguish two types of endowment, that is, direct and indirect. By direct we mean equipment owned by an individual and, with the exception of the cellular phone, installed at the home. We consider as indirect equipment the access that individuals have to telecommunication equipment at their office or with friends, at similar conditions as at home, if not for free. As we believe this to be an important variable, we include it in the analysis of telephone calls, though it will be more interesting once we model adoption and use of new media, such as the internet.

In this first stage, then, the individual has minimized her cost for producing telecommunication by making phone calls and by using her consumption time. For empirical purposes we will want to model the demand for phone calls and not that for telecommunication, however. Hence, applying Shepard's Lemma (Varian, 1992), we take the first derivative of the above cost function which is, under certain conditions not discussed here, equal to the conditional factor demand. In the case of the above general specification we can thus write the demands for phone calls as two functions of the same variables:

$$Q_n = q_n(T, P_d, P_n, \tau_T, R, E_d, E_{id}) \tag{6.3}$$

Behind this cost minimization we have a utility maximization by the individual. She maximizes utility from telecommunications and a Hicksian bundle X. We add personal characteristics Z and attitudes A as taste shifters to the utility function:

$$U = u(T, X, Z, A) \tag{6.4}$$

Maximizing this under the following budget constraints, where Y is income and the price of the Hicksian commodity and the time to consume it, τ_X, are standardized to 1:

$$\begin{aligned} Y &= c(T, P_d, P_n, \tau_T, R, E_d E_{id}) + X \\ \overline{\tau}_C &= \tau_T T + \tau_x X = \tau_T T + X \end{aligned} \tag{6.5}$$

we obtain a demand function for the commodity telecommunication T:

$$T = t(P_d, P_n, \tau_T, \overline{\tau}_C, R, E_d, E_{id}, Z, A, Y) \tag{6.6}$$

Substituting this into (6.3) we obtain the demand function for phone calls as:

$$\begin{aligned} Q_d &= q_d(P_d, P_n, \tau_T, \overline{\tau}_C, R, E_d, E_{id}, Z, A, Y) \\ Q_n &= q_n(P_d, P_n, \tau_T, \overline{\tau}_C, R, E_d, E_{id}, Z, A, Y) \end{aligned} \tag{6.7}$$

This provides the basis for the empirical model we want to estimate. We expect the variables to have the same signs in both equations. The own price should have a negative sign and the cross-price a positive sign. The contact network is expected to have a positive influence on the number of telephone calls, assuming that each individual uses this contact network in an optimal way.

We assume that τ_T depends on the mode of telecommunication but in the empirical model we consider only one mode (that is, telephone). Therefore in this specific context the variable τ_T is fixed and constant.

3. THE DATA

The data used in this analysis were collected by a telephone survey in February 1998. The data-set survey contains data from approximately 650 individuals belonging to households in Ticino and selected from a sample stratified by age, gender and employment status. Information concerning both individual and household level was collected in relation to usage and endowment of traditional and new telecommunication media, employment status, qualification, income and family size as well as psychological attitudes toward new telecommunication systems through telephone interviews. In particular, data on telephone usage were requested, dividing local and long-distance telephone calls both on a weekly basis. The latter were collected in relation to the contact network. This was measured first as the number of people contacted per week by an individual in the two most frequently called places (*primary network*), second by the number of people contacted within Switzerland during a month (*secondary national network*) and third by people contacted within a month outside Switzerland (*secondary international network*). In relation to those two places, the number of calls per week, their average duration and the time of day of the call were recorded.

4. EMPIRICAL MODEL AND ESTIMATION PROCEDURES

From the theoretical model we derive a system of two demand equations for time of day usage (TOD) for long-distance telephone calls. We assume homogeneity between long-distance calls with respect to distance. This means that we treat one minute of phone conversation to London as the same good as one minute to Florence except for the rate,[1] which varies with distance.

We assume that the system demand function is multiplicative and therefore, applying the log transformation, we obtain the semi-logarithmic form of the model to be estimated:

$$Q_{ki} = \alpha'_{k0} + \alpha_{k1} \ln P_{si} + \alpha_{k2} \ln M_i + \alpha_{k3} \ln N_i + \alpha_{k4} \ln NCH_i \\ + \alpha_{k5} \ln NES_i + \alpha_{k6} \ln A_i + \alpha_{k7} E_{id} + \alpha_{k8} CELL_i + \alpha_{k9} FAX_i \\ + \alpha_{k10} INET_i + \alpha_{k11} RET_i + \sum_j \beta_{kj} Z_{ij} + u_{ki} \quad (6.8)$$

where k = day, night/weekend. (The meaning of the variables is explained in Table 6.1.)

In order to explain the procedure for estimating the empirical model, it is necessary to focus on two important methodological aspects. First, we assume that disturbances u_{ki} for night and day call minutes are normally distributed, non-correlated across the observation and correlated across the two equations. Therefore:

$$E[u_d u_n] = \sigma_{dn} l_n = V \qquad d = \text{day};\ n = \text{night/weekend}$$

If disturbances between the equations are correlated, there is a relation between the two equations. In this case, estimating by OLS would yield unbiased but not efficient results. Therefore, efficient estimates could be obtained by using Zellner's SUR model (Seemingly Unrelated Regression model; Zellner, 1979) which consists of generalized least squares applied to the system equation weighed by the V matrix. However, Dwivedi and Srivastava (1979) showed that if the equations have identical explanatory variables, that is $X_n = X_d$, the SUR estimator and OLS applied to the single equation are identical. Therefore in this case, there is no need to adopt SUR estimation, and the same efficiency can be gained estimating the equation separately.

The second aspect is that the data set has an important characteristic: there are several observations where either peak or off-peak calls are zero if an individual allocates all the phone calls in one period. Dependent variables Q_n and Q_d are, in fact, censored variables. This characteristic destroys the linearity assumption of equation (6.7) and LS is clearly inappropriate (Amemiya, 1984). The Tobit ML estimator is, therefore, applied to each single equation to take into account the observations at zero.[2]

The model to be estimated is:

$$\begin{aligned} Q^*_{ki} &= Q_{ki} \quad \text{if } Q_{ki} > \mathbf{0} \qquad k = \text{ day, night/weekend} \\ &= \mathbf{0} \quad\ \ \text{otherwise} \end{aligned} \quad (6.9)$$

where Q_{ki} is the observed variable and Q_{ki} is the dependent variable of equation (6.8). With log-linear variables for the demand functions Q_{ki}, the coefficients estimated (for example for the price index) can be interpreted as follows:

$$\frac{\partial Q_{ki}}{\partial P_s} = \frac{\alpha_{k1}}{P_s} \Rightarrow \alpha_{k1} = \frac{\partial Q_{ki}}{\partial P_s} P_s = \varepsilon_{Ps} Q_{ki}.$$

Table 6.1 Description of the variables

Variable	Description
Q_k	Dependent variable, duration of total calls per week in minutes made during *k* rate period of day.
P_s	Linear approximation of Stone's price index as proposed by Muellbauer and Deaton (1980): $$\ln P_S = \sum_k \overline{w}_k \ln P_k$$ with $\overline{w}_k$ the median expenditure share for long-distance calls and P_k price for a minute of call during day and night period.[a]
$M = m/Y$	Household expenditure for phone calls per month (phone bill, *m*) as a share of income (*Y*).
N	Number of people contacted in the two most frequently called places (primary networks).
NCH	Number of people called in Switzerland with at least a monthly frequency (secondary national network).
NES	Number of people called abroad with at least a monthly frequency (secondary international network).
A	Categorical variable measuring psychological attitudes towards telecommunications.
E_{id}	Dummy variable for indirect endowment that equals 1 for an individual with the possibility of using the telephone without restriction at her work place, 0 otherwise.
CELL, *FAX*, *INET*	Dummy variables for endowment with mobile telephone, facsimile, internet respectively.
RET	Dummy: equals 1 for retired; 0 otherwise. It contains information about consumption time τ_C
Zj *AGE*, *GENDER*, *FOR*	Demographic and socio-economic variables, more specifically: *AGE* as continuous variable ranging from 18 to 65 years; *GENDER* = dummy: equals 1 for female; 0 otherwise; *FOR* = dummy: equals 1 higher education (university degree); 0 otherwise.

Note: [a] The rate also varies as a function of distance. P_k is therefore calculated as a weighted average of the two rates per minute of the calls to the two most-called places. This results in having P_k vary among individuals as a function of distance.

Thus the price elasticity turns out to be:

$$\varepsilon_{P_s} = \frac{\alpha_{k1}}{Q_{ki}} \tag{6.10}$$

In this form, the coefficient of the logarithmic variable can be interpreted as the absolute change of the dependent variable in relation to a percentage change of the independent variable. In contrast, the coefficients of dummy variables explain the impact on the dependent variable in absolute terms:

$$\Delta Q_{ki} = \alpha_{k8}$$

which consists of shift when the dummy equals 1.

5. ESTIMATION RESULTS

In Table 6.2 the MLE of the Tobit demand system obtained by estimating the equations separately are presented. In accordance with the recent telecommunication literature (see Chen and Watters, 1992; Taylor, 1980), total telephone

Table 6.2 MLE parameter estimates of the two rate period equations

Coefficient	*Qn**	*T* value	*Qd**	*T* value
Constant	20.279	(2.106)	6.5195	(0.608)
GENDER	6.9992	(3.166)	5.5885	(2.517)
Ln(*M*)	7.5552	(6.024)	3.6338	(2.915)
Ln(P_s)	–4.9407	(–2.958)	–7.4742	(–4.414)
AGE	–0.0958	(–1.036)	0.0464	(0.508)
Ln(*A*)	4.5992	(2.697)	3.7904	(2.224)
Ln(*N*)	7.3360	(3.969)	4.4437	(2.419)
Ln(*NCH*)	–0.58576	(–0.432)	3.6969	(2.747)
Ln(*NES*)	3.2904	(1.663)	–0.69828	(–0.354)
RET	–7.4579	(–2.060)	–8.2445	(–2.296)
FOR	7.5916	(2.409)	10.822	(3.475)
E_{id}	–6.0476	(–2.322)	–5.8284	(–2.287)
CELL	7.7589	(3.312)	–6.9307	(–2.922)
FAX	2.6956	(0.816)	–2.0914	(–0.618)
INET	2.0098	(1.101)	–0.65610	(–0.359)

Note: * Dependent variables: *n* = night/weekend; *d* = day.

budget share (M) and price have, respectively, positive and negative impacts in both equations. The low-rate period seems to be more sensitive to price variation than the high-rate period (coefficient –4.94 vs –7.47). The opposite holds for the budget share (7.55 vs 3.66).

Table 6.3 presents the means, the dependent variables and the values they take in response to a 1 per cent variation of M and P_s. The findings indicate, for example, that when Q_n = 23.32 min. and the price index varies by 1 per cent we have Q_n = 18.38 min.

Table 6.3 Duration (in minutes) of phone calls when independent variables vary by 1 per cent

1% variation of	Off-peak period $\overline{Q}_n = 23.32$ min.	Peak period $\overline{Q}_d = 16.19$ min.
Expenditure share (M)	30.88 min.	19.83 min.
Price index (P_s)	18.38 min.	8.72 min.

As we expected, the contact network has an impact on both demand functions. The value of the coefficients of the primary contact network is 7.33 and 4.44 for peak and off-peak calls respectively. Secondary national contact network *NCH* has a significant effect only on high-rate period (day) calls, (variation by 1 per cent of this variable has 3.69 minutes variation of the peak phone calls).The opposite holds for the international network *NES* which has an impact on the low rate period (night/weekend) calls. This probably means that individuals call abroad during low-price periods and call inside Switzerland during high-rate periods. In other words, households can choose to wait until the evening or night to make their calls towards people living abroad in order to profit from lower prices. Table 6.4 gives an example of this impact.

The coefficient of indirect endowment variables has the hypothesized negative sign in both equations (–6.04 and –5.82), proving our theoretical expectation. As a consequence, an individual who can have access to phone service at the office without significant restrictions reduces her consumption of long-distance telephone calls.

Mobile phone endowment is found to reduce the demand for peak period calls (–6.9 minutes) and increases the demand for off-peak calls (7 minutes) (see Table 6.5). This seems to indicate that the cellular phone calls are a substitute for home phone calls during the day but have a catalytic effect on home phone calls during off-peak hours. Fax and internet endowment have no significant impact on telephone demand: these media seem so far to be neither substitutes for nor complements to the traditional medium for the average user.

Table 6.4 Impact of contact network on average dependent variable level

1% variation of	Off-peak period $\overline{Q}_n = 23.32$ min.	Peak period $\overline{Q}_d = 16.19$ min.
Primary network (*N*)	30.66 min.	20.64 min.
Secondary national network (*NCH*)	–	19.89 min.
Secondary international network (NES)	26.61 min.	–

Table 6.5 Endowment impact on average dependent variable level

Dummy = 1	Off-peak period $\overline{Q}_n = 23.32$ min.	Peak period $\overline{Q}_d = 16.19$ min.
Indirect endowment (Ei_d)	17.27 min.	10.36 min.
Mobile endowment (*CELL*)	31.08 min.	9.26 min.

Table 6.6 Socio-demographic impact on average dependent variable level

Dummy = 1	Off-peak period $\overline{Q}_n = 23.32$ min.	Peak period $\overline{Q}_d = 16.19$ min.
GENDER (1= female)	30.32 min.	21.78 min.
Retired (*RET*)	15.86 min.	27.02 min.
University degree (*FOR*)	30.91 min.	7.95 min.

Turning to socio-demographic variables (see Table 6.6), the results show accordance with the considerations reported by Taylor (1980): 'calling behaviour cannot be accounted for by age or size of household' and 'phone usage is lower in households headed by a person aged sixty-five or more'. We found gender, retired status and university education to be significant factors. On average, women with these characteristics call 7 and 5 minutes more than men during off-peak and peak period respectively.

6. CONCLUSION

In the research presented we try to explain telephone usage in a general context of telecommunication. The aim is to introduce factors believed to be relevant for the adoption of new media (contact network, direct endowment with alternative media and indirect endowment with the same medium) in a state-of-the-art model of telephone demand. Household production theory was used as a background for formulating the demand functions that were then estimated using the Tobit approach. The results confirm the findings of other studies as far as 'standard' variables are concerned.

In relation to the contact network we found that its size in general and its location inside and outside of Switzerland have a significant impact on telephone demand and its allocation to peak and off-peak periods. We conclude that the interaction between long-term contact decisions (investments in links) and short-term communication behaviour has to be understood if the adoption and use of new media is to be analysed.

The indirect endowment introduced in this study also has the expected relevance for telephone demand. Opportunities to access the same medium (in our case phone) at the work place or with friends indeed reduce the demand for its use at home. This, we conclude, will also be of relevance for adoption and use of, for example the internet, and if modelled correctly it will lead to more precise forecasts of the adoption of this medium.

The results obtained with respect to direct endowment with alternative media are also of significant interest. Fax and internet endowment and use have no significant impact on telephone demand: these media seem so far to be neither substitutes for nor complements to the traditional medium for the average user. In the case of the cellular phone, it is found that it reduces (*ceteris paribus*) the demand for peak period calls and increases the demand for off-peak calls. This seems to indicate that cellular phone calls are a substitute for home phone calls during the day but have a catalytic effect on home phone calls during off-peak hours. Without further evidence it is difficult to make more detailed hypotheses, but in any case the cellular phone seems to be a substitution of home calls during the day when people tend to be out of the home, and to have a reinforcement effect during the night rate period. This latter effect, expected to be of relevance also for other media by many experts, seems a key issue for future work.

The next steps planned in this study will therefore include the other media, and given the context of low usage and highly specific behaviour, the application of stated preference analysis is foreseen.

ACKNOWLEDGEMENT

This research is part of the project 'Switzerland towards the Future: New Media, Communication Network and Social Change', funded by Swiss National Research. The authors acknowledge the comments of anonymous referees on an earlier draft and Gina Poncini for the last review of the English version.

NOTES

1. Swisscom offers two rate periods for international calls: day (8:00–21:00, Monday–Friday), night/weekend (21:00–8:00 Monday-Friday and 0:00–24:00 on Sunday and Saturday) and three rate periods for national calls: day (8:00–17:00, 19:00–21:00, Monday-Friday); morning/evening (6:00–8:00, 17:00–19:00, 21:00–23:00, Monday-Friday); and night/weekend (23:00– a.m. all days).

   For comparability reasons the three rate periods have been transformed to two rate periods. The morning/evening price has been considered as day rate for the morning hours and night rate for the evening hours.
2. For the model $\begin{cases} Q_i^* = X'\alpha + \varepsilon_i \\ \quad = 0 \end{cases}$ the likelihood function to be maximized is

$$L = \prod_0 (1 - \Phi_i) \prod_1 \frac{1}{|\Sigma|^{-1/2} (2\pi)^{n/2}} e^{-1/2(Q_i - \alpha' x_i)\Sigma^{-1}(Q_i - a' x_i)}$$

where the first product is calculated for the zero observations and second one for the others, assuming normality distribution and

$$\Phi_i = \Phi(\alpha' x_i, \sigma^2) = \int_{-\infty}^{a' x_i} \frac{1}{|\Sigma|^{-1/2} (2\pi)^{n/2}} e^{-t^2 \Sigma^{-2}} dt \text{ is a density function calculate in } \alpha' x_i / \Sigma$$

others, assuming a normal distribution, and

$$\Phi_i = \Phi(\alpha' x_i, \sigma^2) = \int_{-\infty}^{\alpha' x_i} \frac{1}{|\Sigma|^{-1/2} (2\pi)^{n/2}} e^{-t^2 \Sigma^{-2}} dt \text{ is a density function calculated in } \alpha' \mathrm{x}_i / \Sigma.$$

REFERENCES

Amemiya, T. (1984), 'Tobit Model: A Survey', *Journal of Econometrics*, **24**, 3–61.

Baker, G.S. (1965), 'A Theory of the Allocation of Time', *The Economic Journal*, **75**(299).

Chen, A. and J.S. Watters (1992), 'Estimating Telephone Elasticities: A Share Equation System Approach', *Applied Economics*, **24**, 1219–24.

Dwivedi, T. and K. Srivastava (1979), 'Optimality of Least Squares in the Seemingly Unrelated Regression Model', *Journal of Econometrics*, **7**, 391–5.

Epple, D. (1987), 'Hedonic Prices and Implicit Markets: Estimating Demand and Supply Functions for Differentiated Products', *Journal of Political Economy*, **95**, 59–80.

Filippini, M. (1995), 'Swiss Residential Demand for Electricity by Time-of-use', *Resource and Energy Economics*, **17**, 281–290.

Muellbauer, J. (1974), 'Household Production Theory, Quality and Hedonic Technique', *American Economic Review*, **64**, 977–994.

Muellbauer, J. and Deaton (1980), 'Economics and Consumer Behaviour', Cambridge: Cambridge University Press.

Taylor, D.L. (1980), *Telecommunication Demand: A Survey and Critique*, Cambridge, MA: Ballinger Publishing Company.
Varian, Hal R. (1992), *Microeconomic Analysis*, New York: W.W. Norton & Co.
Zellner, A. (1979), 'An Efficient Method of Estimating Seemingly Unrelated Regressions and Tests of Aggregation Bias', *Journal of the American Statistical Association*, **57**, 348–68.

PART III

Traffic and Congestion

7. Infrastructure and congestion: Can rail save the road? Can public transport replace the car?

Piet H.L. Bovy and Bert Van Wee

1. INTRODUCTION

Road traffic congestion is a widespread phenomenon in affluent societies. This chapter will address the recurrent congestion mainly caused by urban commuters on motorways and other trunk roads at the fringes of the larger high-density conurbations. Reducing the congestion burden and its related safety and environmental problems is a general concern to people and therefore to policy makers at the national governmental level.

Most transport experts, for example Hillman (1993), and critics of current transport policy continue to believe that the key to resolving the conflict between the growth in the demand for car travel and its adverse effects (such as congestion, air and noise pollution, lack of safety and so on) lies in investing heavily in bus and rail rather than roads to provide equivalent levels of comfort, service and speed to the car. In this way, it is thought that car and lorry users can then be more easily encouraged to transfer to public transport or rail transport, or even be obliged to do so with fewer grounds for opposing measures taken by central or local government with this aim in mind. Such judgements, well substantiated by Hillman (1993), can be seen at nearly all levels of government, in most spheres of information gathering and dissemination, in professional papers, political lobbying and agenda setting, and thus not surprisingly in the media. In Hillman (1993) an overwhelming account of examples of such propositions is given.

The purpose of this chapter is to elaborate on these beliefs and to present some non-traditional viewpoints on road traffic congestion. If we accept the generally shared view that we suffer from road traffic congestion and that we need policies to reduce the congestion burden (at least because of economic reasons), then we can make a few statements about the relative value of investing in the road system or in the public transport system to curb congestion.

The discussion is structured around three assertions or statements with respect to congestion policy:

1. 'Congestion is not necessarily a sign of a bad network' (or of a bad transportation system). It is more likely a sign of a good network. Perhaps the roads are too good. A number of paradoxes appear to emerge if we look at congestion developments in economically well-developed regions.
2. 'Only the road can relieve the road.' This quote from the lucid book by Gerondeau (1997) suggests that only interventions in the road traffic system and vehicles can effectively reduce road problems. It argues that public transport cannot solve road traffic problems and is too costly and ineffective.
3. 'Environmental gains of rail and public transport investments are often absent or even negative.' Most proposed rail and public transport investments do not contribute to congestion relief or environmental improvement, they are even detrimental to environmental objectives. They may, however, serve other purposes effectively.

Such viewpoints are raised only very seldom, for example in Hillman (1993), Gerondeau (1997), and Rubin et al. (1999). The opposite views where public transport or rail investments are proposed as solutions to our transportation problems are heard much more loudly and frequently.

These three statements are crude, not always valid for specific cases, and are provocative. Nevertheless they are useful for clarification of important planning issues and may well serve the ongoing discussion and policy debate. Equally, though, it is a Western European perspective, but the arguments and insights may well be of value for and applicable to developing economies, such as Israel or the Eastern European countries where lively policy debates surround transportation investment strategies. The recent public debate about the construction of Highway 6 as a major north–south corridor by-passing Tel Aviv, or, alternatively, developing rail serves as a good example of this general issue.

It is our view that in assessing policies to improve transportation conditions the basic objective is to maximize the net utility associated with travelling and transporting under the restrictions of sustainability (environment, safety). The next sections in turn elaborate on one of the assertions formulated in the introduction. Since road congestion as the central issue of this chapter is being caused by passenger as well as freight transportation on the road, we will look at the role transport alternatives offered by public transportation on the one hand (bus and rail based) and rail freight transport on the other can play in alleviating the congestion burden and environmental pollution.

Although they are very powerful alternatives to car use, this chapter will not discuss the potential contributions of walking and cycling. For an elaborate discussion of the contributions of these modes, see for example Hillman (1993).

2. TRAFFIC CONGESTION: NOT NECESSARILY A SIGN OF A BAD NETWORK

Traffic congestion, be it on the road, in the railway stations or in airports, is a real burden to the individual and to society at large. Travelling becomes highly unreliable, with high travel time losses, extra fuel consumption and emissions.[1] The report on the 110th Round Table of the ECMT gives an account of the road congestion situation in Europe, with national reports on the UK, France, Germany, and the Netherlands (ECMT, 1999). Although a clear, unambiguous and widely accepted definition of congestion does not yet exist, nor accepted standards for measuring the prevailing level of congestion, the general belief, at least of policy makers and in the media, is that current levels of road traffic congestion in Europe exceed acceptable levels and that something should be done.

Based on spatial analyses, road traffic congestion appears to be a limited phenomenon, concentrated at the periphery of the larger, densely populated conurbations (Bovy and Salomon, 1999). Less than 10 per cent of the total population of European commuters suffers from daily congestion (Gerondeau, 1997). The majority of commuters succeed in keeping travel between home and work around 20 minutes or less.

Despite the fact that traffic congestion appears to be a highly local problem for only a small part of the travelling population, the question remains to be answered whether this situation is unacceptable and should be remedied.

Some European policy documents state that the costs of extra travel time associated with current levels of congestion amount to about 2 per cent of GDP (Kinnock, 1995).[2] Although the validity of this cost figure might be questioned (see Bovy and Salomon, 1999), even if it were only one-quarter of this level, which seems more realistic, these costs in an absolute sense are tremendous. Nevertheless, we will argue that it would be foolish to try to get rid of all congestion, as is often heard in policy discussions, and to strive for zero congestion levels (see also Rubin et al., 1999). Trying to eliminate congestion altogether, regardless of whether this would be possible, of course also costs huge sums. An excessive transport network would be needed to that end, with enormous excess capacity during most of the time.

Somewhere there is an equilibrium between the costs of measures to reduce congestion (for example building more roads) and the costs of congestion itself. The reduction in congestion costs due to adding measures beyond this point is lower than the costs of the additional measures. There exists, at some point, an optimum level of congestion which can be calculated given the cost figures for road construction, road maintenance, safety, environmental damage and congestion impacts. This optimum level depends on the contextual situation.

In the Netherlands economic analyses have shown that (in 1990) this optimum for the Dutch trunk road network (including all motorways) is on average at about 2 per cent congestion probability (Stembord, 1991). This congestion probability measure relates to a section of road and means that over a longer period 2 per cent of daily traffic on that road section will encounter congestion to some degree on working days. For example, if we assume that all traffic congestion concentrates in the morning and afternoon peak hours (two hours per day), each of which carries about 10 per cent of daily traffic, on average 10 per cent of daily peak hour travellers will experience congestion (queues). That is only a few minutes delay in a queue once a fortnight. Therefore, if it is not too bad, and if its level is not too high, congestion is a sign of an efficient, well-designed network.

Furthermore, it appears that in many cases congestion is a sign of an affluent, abundant network with too many motorways, too many access points, too high a spatial density and motorway access densities such that nearly everyone lives close to a motorway entry. Using the Netherlands for illustration, one finds that the Dutch motorways constitute 2 per cent of the total national road network (measured in length) but about 80 per cent of the population and jobs are within ten minutes from a motorway access point. Other European regions show similar characteristics (see for example Bovy, 2000). Therefore about 40 per cent of all car kilometres are driven on motorways, which account for only 2 per cent of the transport network. On average, there is an entrance (and an exit) every eight kilometres of motorway and, in the highly urbanized Randstad region, every four kilometres. The motorway network is so attractive that nearly everyone wants to use it. Despite congestion due to the high demand, the motorway is seen as the best alternative: better than using local or regional roads, better than using bus or train in most cases.

Congestion is not necessarily a sign of bad or worsened conditions for the traveller. As one survey respondent noted: 'I now have to stay ten minutes in a jam every day, but a few years ago this trip took me half an hour longer' (GoudappelCoffeng, 1997). Despite current widespread congestion, interurban travel times in general have not increased. Instead, they have decreased strongly in the last 20 years, and are still decreasing slightly. Meanwhile, average interlocal travel speeds have increased by about 1 per cent per year (CBS, 1999). This observation, also made in the United States (Gordon and

Richardson, 1991), seems paradoxical and needs further explanation. The spatial shifts in activity patterns in the last decades have resulted in an increasing share of longer intercity trips, especially commutes, and a reduction of short-distance intra-urban trips. Given the high density and accessibility of motorways, steadily increasing portions of these intercity commutes are performed on high-speed motorways. This activity contributes significantly to the traffic loads and thus to congestion there. The use of low-speed urban roads is gradually being substituted by high-speed motorways. Much longer distances are covered at higher speeds, resulting in equal or even decreasing travel times. While these shifts contribute to increasing congestion, the net effect for many travellers is a constant or even slightly reduced travel time.

Not only travel times are important. The reliability of the travel times for given trips is also important for users of the transport system (Peeters et al., 1998).

Constant or even improved interurban door-to-door travel speeds are not the only paradox. While the statistics clearly reveal that the total level of congestion in a region steadily increases (for example in terms of total hours spent in queues, or total hours lost, or total congestion cost), the individual drivers do not experience, on average, an increase in congestion loss. Time lost in queues appears to be about constant at the individual level (Transpute, 2000). This paradox can be explained to a large extent. The aggregate growth of congestion stems from the increasing travel demands that are not sufficiently matched by corresponding increases in capacity. This growing demand stems from population growth, employment growth, increased participation of women in the workforce, and other autonomous factors. The resulting traffic levels on the motorways lead to an increased number of recurrent daily jams and especially a longer duration of jams. Jams start earlier in the morning and take longer during the day. However, queue lengths do not change much. As a result, the time lost in a queue by the individual driver also does not change much. The increase in congestion indicators therefore is not the result of an increase of travel times of each individual vehicle user, but of an increase in the number of individuals facing congestion. It is hypothesized that travellers accept a certain maximum time loss due to congestion, calculated to be roughly in the order of about 12 to 15 minutes per trip in the Netherlands (Transpute, 2000). Of course, this maximum is specific to the individual. If this maximum level is reached, the traveller looks for alternatives, most frequently by travelling earlier or later, or choosing a different, probably longer, route. Switching to another mode of transport is seldom selected. This equilibration mechanism leads to a spreading of congestion patterns in space and time. Note that the overall congestion level expressed as 'lost vehicle hours' increases, but the number of vehicle kilometres also increases. For example, between 1990 and 1997 in the Netherlands the number of lost vehicle hours increased by 35 per cent (NEA, 1998), but the number of road traffic vehicle kilometres on motorways increased by 39 per cent in this period. So, relatively speaking the congestion level did not increase at all.

To summarize, congestion does not necessarily point to a badly designed transport network. Efficient networks exhibit a certain optimal level of congestion at which optimal use is made of available resources. In many cases, the access characteristics of the network are so attractive that the resulting local congestion for the travellers is outweighed by the high door-to-door level of service. The congestion paradoxes show that it is necessary to distinguish clearly between aggregate congestion measured at the level of the network and the individually experienced congestion measured at the level of the traveller.

3. 'ONLY THE ROAD CAN SAVE THE ROAD'

The second point for discussion is how to address congestion that is considered unacceptably high. Similar questions have been raised regarding other problems with road traffic, such as safety and environmental damage. In a lucid treatment of these questions, Gerondeau (1997) summarized the only effective approach to solving these problems with the assertion: 'Only the road can relieve the road'. This assertion suggests that the only really cost-effective way to reduce road traffic congestion or improve environmental quality is to intervene directly in the road traffic system itself, such as by building (better) roads, regulating access, levying tolls, and the like. It does not make sense to improve the alternatives to the car for that purpose. Such alternative policies always turn out to be costly and ineffective detours on the way to a successful solution. Improving public transport to reduce congestion or environmental impacts of transportation is a particularly costly and ineffective approach for a number of reasons.

First, there is the extreme imbalance between car use and public transport use. The Netherlands certainly is a representative country to illustrate this aspect. In the Netherlands about 60 per cent of all personal trips are made with the car, compared to only about 6 per cent with public transport. As a result, there is a car/PT (public transport) trip ratio of 10 to 1 (CBS, 1999). In terms of passenger-kilometres travelled this ratio still is 7 to 1 because of the relatively long public transport trips. For similar figures for the UK, see Hillman (1993). So, even if there were a perfect substitution between car and public transport, the public transport system would have to be improved and enlarged enormously in order to reduce car use noticeably. Increasing the use of existing systems (for example more trains on the same rails) might result in lower marginal costs compared to average costs (costs per passenger-kilometre), but even then the cost-effectiveness will very likely be poor, at least if current subsidies on fares and building infrastructure remain the same.

However, the substitution between car and public transport is fairly weak. The cross-elasticities are fairly low both with respect to prices and travel times. Table 7.1 shows a collection of direct and cross-elasticities of demand established for the Netherlands (Van der Waard, 1990).

Table 7.1 Long-term elasticities of demand (passenger-kilometres travelled) with respect to time and costs of car and public transport respectively

	Travel time			Fuel price	PT fare	
	Car	Train	Other PT	Car	Train	Other PT
Car driver	–1.3	+0.03	+0.05	–0.2	+0.02	+0.02
Car passenger	–0.7	+0.03	+0.08	+0.15	+0.02	+0.05
Train use	+0.5	–1.6	–0.33	+0.14	–0.8	–0.04
PT big cities	+0.34	–0.03	–1.22	+0.14	–0.03	–0.65
Other PT use	+0.45	–0.12	–1.54	+0.15	–0.06	–0.7

Source: Van der Waard (1990).

The effects of car costs on public transport use appear to be relatively small. The cross-elasticity of public transport kilometrage for the fuel costs is +0.14, averaged over all travel purposes and all types of spatial relations (with and without good public transport). Comparable figures hold when the market segments to which they apply are taken into account (Bovy et al., 1991). Long-distance rail trips (>40 km) show a somewhat higher value of +0.22, while the non-captives in commuting appear to be a bit more sensitive: +0.26. These two higher elasticities are due to the fact that within these specific segments the availability of public transport alternatives is higher.

The cross-elasticity of travel time by PT on car use appears to be very low (about +0.03). For longer trips (>40 km) a value of +0.17 applies, which is comparable taking into account the difference in market segments. Over longer distances, the train competes very well with the car with respect to travel time, and therefore shows a substantial modal share of about 15 to 30 per cent. However, the largest contingent of all trips takes place on the short and medium distance range, where public transport does not (yet) offer a competitive travel time (see Table 7.2).

Table 7.2 Passenger transport per mode, passenger-kilometers $\times 10^9$, 1998

Car driver	93.8
Car passenger	55.2
Train	14.9
Bus/tram/metro	6.3
Other	29.7

(Cross-)elasticity values do not sufficiently show the shifts between markets. To show the size of shifts, elasticities have to be combined with the volume by mode. Table 7.2 provides data on the volumes for the Netherlands, 1998.

The low overlap in markets can be illustrated by the following example. In the Netherlands the number of car-driver-kilometres is about six times as high as the number of passenger-kilometres by train (see Table 7.2). Suppose the government wants to reduce travel times by train in order to persuade car drivers to switch to train. A 1 per cent reduction in travel times by train results in an increase in passenger-kilometers by train of 1.6 per cent and a decrease in kilometres of car drivers of 0.03 per cent (Table 7.1). To compare absolute changes in kilometres we can express this latter percentage in passenger-kilometers by train, multiplying it by six, resulting in a share of approximately 0.20. In other words, the increase in train use is about eight times higher than the decrease in car use. The same calculation may be made with respect to a decrease in train fares (for example by additional subsidies). Then the increase in train use is about seven times as high as the decrease in car use.

It is generally accepted that a realistic substitution potential from car to PT exists if the door-to-door travel time ratio between PT and car is less than about 1.5. Table 7.3 shows the current PT/car travel time ratios for different distance classes as well as the current shares of car trips and PT trips in each distance class. Only at distances over 30 kilometres does PT appear competitive.

Changes in public transport fares appear to have only a limited effect on the substitution of travel demand between car and PT, as is shown in many studies, surveys and model estimates (Bovy et al., 1991). An explanation for this is that changes in costs hardly affect the availability of the different transport modes, as is for instance the case with changes in travel times. Further explanations relate to the high weight travellers apparently assign to quality, the low public transport fares relative to income, and the fact that the costs of business and most commuting trips often are paid by the employer. The price elasticity of PT (train) to car use is about +0.02 on average, while

Table 7.3 Door-to-door travel time ratio by distance class with shares of car trips and PT modal shares

Distance band (km)	Average tt-ratio PT/car	Share of car trips (%)	PT modal share
0–5	5.1	38	1
5–10	4.3	20	6
10–15	3.7	11	8
15–30	2.5	18	11
30–50	1.3	7	14
>50	1.1	6	20

Source: Bovy et al. (1991).

over longer distances (>40 km) this is about doubled. Commuting is more sensitive to changes than other travel purposes; and non-captives in commuting again prove to be twice as sensitive (+0.28) to changes in PT fares. However, the more sensitive segments of the travelling population form only a small percentage of the total.

The reasons for these low cross-elasticity values deserve some elaboration at this point. First, public transport is a viable travel alternative only for a small part (roughly 25 per cent) of the total travel market. This has apparently been the case at least during the last four decades and will most probably remain the case in future. By its very nature, a public transport system will always have a coarse network with a low access density and relatively low speeds at short and medium distances. At long distances, public transport (mostly by train) can be competitive, but the travel market is very small and has thin flows. Thus the spatial coverage of travel demands by public transport is strongly limited and mostly poor, except for high-density conurbations. The fundamental difference between the rail and road transport systems is the spatial and temporal coverage of travel demand, which limits the substitution between these modes.

Various empirical studies (for example RUA corridor, TNO–Inro, 1998) have shown that rail mainly facilitates trips with origins and destinations located in the served corridors. The catchment area of public transport services is limited to the direct vicinity of the stations (among other reasons because of the high impact of transfers). In contrast, roads facilitate trips with very dispersed patterns of origins and destinations and with a nearly continuous catchment area around the road facilities (see Gerondeau, 1997, p. 330, fig. 9). In addition, numerous empirical studies suggest that new rail connections attract travellers from other public transport services. The share of former car users is always very small, in the order of magnitude of 25 per cent of rail passengers (Bovy et al., 1991).

Travel choices are also limited by captivity of the travellers. About half of the car users are captive to the car, as alternatives cannot meet their needs. On the other hand, the low substitution from car to public transport is also due to the many other choices a car user has available if he/she is confronted with extra burdens in time or cost:

- other modes: bicycle, ride sharing such as in the case of introducing tolls or other measures;
- other departure times: to compensate for increasing congestion levels or time-varying tolls;
- other destinations: to compensate for increasing parking rates or other travel costs;
- other car types: to compensate for higher fuel or car taxes;
- higher compensation from employers for increasing travel costs.
- a reduction in the number of trips (four working days per week instead of five; not going home for lunch), to compensate for an increase in travel times or costs.

Only in a minority of cases will the car user switch to public transport if he/she encounters a deterioration in car level of service. Generally, non-PT options will be considered first.

To summarize, given the mobility statistics and the findings of empirical studies, the prospects of short-distance (say, 7 kilometres or less) PT bus and other modes playing a significant role as substitutes for car journeys are limited. These findings also suggest that, for journeys over 7 kilometres, the prospects of a significant diversion of journeys from car to PT by rail are also small. On the basis of current travel patterns by car, the expectation that public transport in general, and rail in particular, can provide a realistic alternative, given sufficient improvement in the quality of its service, is most likely to remain unfulfilled. The car has generated patterns of spatio-temporal activities which cannot conveniently, sensibly, or viably be served by public transport, especially not rail.

It should be made clear that current levels of car use are not at all the result of former transfers from public transport. Historical travel statistics clearly show that this cannot be the case (see for example Hillman, 1993). The growth in car travel has not emerged from shifts from public transport, but from new patterns of activities generated by automobile availability. The belief that improved public transport, especially rail, would re-attract car users neglects the fact that the majority of these car trips were not previously made by public transport.

In short, public transport improvements cannot effectively solve road traffic problems. Direct interventions in the road system are required effectively to reduce congestion and environmental impacts caused by road traffic. Investments in public transport should be assessed on its ability to provide mobility to various groups in society, as well as its ability to support desired spatial organization of our cities.

4. PT INVESTMENTS CAN BE DETRIMENTAL TO THE ENVIRONMENT

The third above-mentioned assertion was: 'Many proposed rail or, generally speaking, public transport investments, don't contribute to environmental improvements. Conversely, they are most often even detrimental to environmental quality.'

The following four reasons contribute to this disappointing outcome. First, the environmental impacts (CO_2, NO_x, energy, noise, and so on) of PT modes (bus, tram, train) in producing traveller-kilometres appear much less favourable compared to other modes than is suggested by many published statistics.

Second, shifts from the car to PT resulting from PT improvements or investments are generally very low (see previous section).

Third, in contrast to the limited car-to-PT diversion, shifts from environmentally friendly modes such as walking and cycling to PT are often large, especially at the lower distance scale and in urban conditions.

Fourth, PT improvements often induce new travel demands (more frequent trips, longer trips), giving rise to additional environmental burdens.

Policies to encourage public transport, especially those involving new rail connections, depart from the assumption that rail transportation exhibits much less environmental burdens than passenger transport by car, and to a lesser extent by bus. These assumptions, however, appear to be strongly exaggerated and cannot sustain a thorough empirical analysis.

A close examination of the relative performance of the various transport modes shows that the environmental achievements of rail transport are not superior to the other modes in all circumstances. In addition, the cost-effectiveness of measures taken in the rail system to reduce the environmental burden (especially emissions) of the transportation sector appears to be very poor. This result is due to the very high investments required and very low realization of shifts from road to rail.

What then is wrong with all these rail proposals? Let us confine the discussion to the environmental performance indicators of the various modes. This approach thus leaves aside the problems of estimating demand levels (mostly far too high) and cost levels (mostly far too low) for rail projects. In developing plans in the public transport sector, it is still common to apply flawed indicators of environmental performance and to compare apples and oranges. In comparing modal alternatives, the emission characteristics of distinct modes used (for example expressed in mass units of emittants per passenger-kilometre travelled) are typically derived from the ratio of total energy consumption or total emission of a mode to total passenger-kilometres produced by that mode over a certain period of time. We may call this the total average method.

Based on these figures, it appears that the environmental performance of rail travel (in terms of, for example, emissions of CO_2 and NO_x) per passenger-kilometer is at least two times better than car travel – see Table 7.4.

Table 7.4 Emission factors (g/passenger-km) of car and train, the Netherlands, 1995

	Car	Train
CO_2	129	63
NO_x	0.78	0.14
CO	2.75	0.01
VOC	0.43	0.01
SO_2	0.09	0.03
PM	0.03	0.00

Note: VOC = volatile organic compounds; PM = particulate matter.

Source: Van den Brink and Van Wee (1997).

In this total averaging approach no distinction is made between spatial conditions, travel distances, vehicle types, competing submarkets, and the like. However, in determining the relative performance of rail (which is only a viable transport alternative at longer distances) it does not make sense to include also short- or medium-distance travel, which is nearly exclusively made by cars or bus alone. The relatively low environmental performance of the car is caused in large part by the overwhelming share of short travel distances in urban conditions that are unsuited for transport by heavy rail.

A valid comparison of environmental performance of modes should be based on a distinction of market segments where the considered modes really can compete with each other and on the specific modal performances achieved in each of these markets. Such a comparison should take account of the use of specific vehicle types (size, propulsion) in each submarket (cars, buses, trams in cities, heavy rail and car in long-distance traffic). Market-specific performance figures are much less favourable for rail than simple average statistics (Janse et al., 2000). For passenger transport this is less clear. A regional train, making many stops, has a higher than average energy use per passenger or seat kilometre, but cars also use more energy per kilometre for regional trips than for long-distance trips.

In many rail investment studies the future environmental performance of rail vehicles (for example electrification) is taken into account, but often corresponding future developments in road vehicles (cars, buses) are neglected. The reason for this is that such studies often consider emission factors for future years, assuming current policy, for example, with respect to emission standards. The effects of new emission standards for which there is EU agreement are taken into account, but additional legislation is not. Therefore such studies may be conservative with respect to long-term developments in road vehicle technology that will result from future legislation. This is more important for road vehicles than for rail, because currently no emission standards for diesel trains exist and there is also no EU (or national) legislation for electric trains.

Environmental gains from public transport projects are generally negligible or even negative in some cases (CPB, 1998; Janse et al., 2000). The reason for this is that for some submarkets the difference in environmental performance of rail and road appears smaller than might be suggested by the main average statistics. This implies that savings from using public transport modes instead of 'bad' motorized individual modes, assuming constant origin–destination patterns, are small. Besides, to compare overall environmental impacts, we have to include the effect of induced demand. The increase in public transport use due to lower travel times or tariffs is about seven to eight times the decrease in car use (see section 3).

To illustrate the environmental relevance of these figures consider the following example. In the Netherlands the energy use of one car-kilometre is about 2.5 times higher than the energy use of one train-passenger-kilometre. Therefore reducing the overall train fares by, for example, a subsidy or increasing the speed of trains will result in a net *increase* of energy use for passenger transport. If there are any savings due to substitution, this does not compensate sufficiently for the high level of generated (induced) travel demand due to higher speeds or shorter travel times. This extra demand, being often the result of substitution from 'green' modes (train, bus, bicycle, foot), naturally induces extra environmental burdens. This shows that the mere fact of stressing public transport patronage growth figures or of full trains and buses as such is not a sufficient indicator for improvements of the environmental burdens of passenger transport as a whole.

When we combine the low effectiveness in reducing road traffic congestion and reducing environmental burdens from traffic with the very high costs of PT services and investments, it is no surprise that the overall cost-effectiveness of PT and rail projects appears very low (CPB, 1998). These unfavourable conditions have been demonstrated in a number of Dutch case studies (HSL-South, CPB, 1998, Student Travel Pass, and the like).

Two concluding remarks are in order. First, it should be noted that the environmental impact of rail investments and a shift from road to rail varies significantly between countries, depending on the way electricity is generated. In Norway the environmental performance of rail is very high because all energy in Norway is produced by hydro-electric power plants. In France and Belgium the share of nuclear energy is much higher than in the Netherlands or Germany. In this chapter we do not discuss the disadvantages of nuclear energy but it should be noted that direct CO_2 emissions of nuclear energy are zero. Second, if a substitution from road to rail should be the result of reduced travel times due to higher speeds of trains, an additional negative effect occurs, as energy use of trains increases with speed. A regular Dutch train which connects the major cities and towns consumes 8 per cent more energy if it drives at 160 km/h than if it drives at 140 km/h (Gijsen, 2000).

5. GOODS TRANSPORTATION BY RAIL

As in the passenger transport sector, there is also a strong belief that increasing the role of rail in goods transportation is always beneficial from an environmental and societal point of view. As will be discussed below, this assumption is highly questionable.

In many countries, as well as at the European level, encouraging rail freight transport is a major issue. Enormous investments are at stake as well as the costs of affecting environmental resources. This raises the questions of whether it is it always wise to encourage freight transport by rail, and whether we should intensify rail freight services. Most debated is the question of constructing new rail lines dedicated solely to freight transportation. The European Commission strongly advocates the establishment of a European network of high-capacity rail freight links. Such projects are under study, for example, in a new Alps Railway Crossing (Alp Transversals) connecting Switzerland and Austria with Italy, a new freight rail link connecting Rotterdam to Western Germany, and high-speed rail freight connections in France. As a part of this European freight transport policy, road freight movements are to be taxed in order to achieve a sensible shift from road to rail. These plans assume that rail freight transport incurs much less environmental cost than goods transportation by road and to a lesser extent by waterway.

These assumptions, however, appear exaggerated and do not bear rigorous empirical analysis. A closer examination of the relative performance of various transport modes shows that the environmental achievements of rail freight transport are not at all superior to the other modes in most circumstances. In addition, the cost-effectiveness of measures taken in the rail system to reduce the environmental burden (especially emissions) of the goods movement sector appear very poor.

What then is wrong with all these rail plans? Let us confine the discussion to the environmental performance indicators of the various modes. As in the case of passenger transport, this approach leaves aside the problems of estimating demand levels (mostly far too high) and cost levels (mostly far too low) for freight rail projects.

In the freight sector, it is still common to apply flawed indicators of environmental performance and to make the wrong modal comparisons.

As in the case of passenger transport, the same total average method is used to calculate emissions for modal comparison. Based on these figures, it appears that the environmental performance of rail freight transport (in terms of emissions of CO_2 and NO_x) is about five or even ten times better than road freight. Mode-specific energy consumption (in megajoules per ton-kilometre) nowadays in the Netherlands is (CE, 1997, p. 38):

Van	19.3
Light truck (< 7 ton)	4.3
Medium truck (7 –18 ton)	2.2
Heavy truck (> 18 ton)	1.8
Train	0.4–0.8
Barge	0.4–0.5

In this total averaging approach no distinction is made between goods types, transport distances, vehicle types, competing submarkets or the like. However, in determining the relative performance of rail (which is only a viable transport alternative at longer distances and for heavy goods) it does not make sense to include short-distance transport, which is nearly exclusively done by trucks. The relatively low environmental performance of freight transport by trucks is caused, among others things, by the bulk of short transport distances that are unsuited for transport by rail or barge.

A valid comparison of environmental performance of modes therefore should be based on a distinction of markets where the considered modes really compete with each other and on the specific modal performances achieved in each of these markets. Such a comparison should take account of the use of specific vehicle types (size, propulsion) in each submarket (light trucks in city distribution, heavy articulated trucks in long-distance traffic). In turns out that market-specific performance figures are much less favourable for rail than simple average statistics (Rutten, 1995). Also, in studies comparing future environmental performance of different modes for goods transport, improvements in future performance of trucks are sometimes underestimated, for the same reason as for passenger transport (see for example Commissie Betuweroute, 1995). Recent developments, enforced by (future) European legislations, show that major improvements in emission performance can be observed (and will be realized in the near future) in the road transport (and in the long term, also the barge) sector, whereas the rail sector is stagnating in this respect. Electric trains benefit from improvements in power generation. In the next decades improvements in diesel trains will very likely be limited, first because no emissions regulations are foreseen, and second because diesel locomotives last much longer than trucks, resulting in a slow penetration of new techniques in the vehicle fleets.

In the course of a policy analysis performed for the planned Betuwe rail freight line from Rotterdam to Germany, market-segment-specific environmental performance figures have been established for the various alternative modes (Janse et al., 2000). It concerns the medium-distance (150 km) markets for point-to-point bulk transport and intermodal container transport. It appears that the estimated bands of emission figures by mode and transport type strongly overlap, so that a clear statement about which mode performs best cannot be given beforehand. Table 7.5 shows representative results of the derived performance ratios for the modes rail, heavy truck and inland navigation with respect to CO_2 and NO_x emissions per ton-kilometre.

With respect to energy consumption and CO_2 emissions, it turned out that the three relevant modes in the two distinguished markets perform compara-

Table 7.5 Indicative CO_2 and NO_x emission figures of road and barge transport relative to rail for the medium-distance markets of bulk and intermodal container transport, 2010

Bulk goods	Intermodal	Container
CO_2		
road/rail	2	1.0
barge/rail	1	0.8
NO_x		
road/rail (diesel)	1	0.5
barge/rail (diesel)	1	0.5
road/rail (electr.)	10	5.0
barge/rail (electr.)	10	5.0

Source: Adapted from Janse et al. (2000).

bly. Only if electric locomotives are used does rail show a clear advantage, mainly with respect to NO_x, but also with respect to CO_2. However, the use of electric propulsion by the rail freight companies is highly unlikely given the expected market conditions (costs, required flexibility, cross-border operations, and so on).

Given these relative figures, investments in rail in order to reduce environmental costs stemming from goods transport are not very cost-effective compared to other measures. On the contrary, the cost of reducing one unit of emission caused by goods transport by investing in rail infrastructure is very high, at least ten times as high as with other measures in the transport sector (Janse et al., 2000). With respect to the planned dedicated rail freight line to Germany, the estimated contribution of this 150 km double-track line to the intended reduction of emissions of the goods transport sector appeared very small (less than 2 per cent of total emissions in the Netherlands due to the transport of goods) and might even be negative. The impact on congestion is also very limited (Janse et al., 2000).

In conclusion, from the case study in the Netherlands, it appears that rail is not at all a mode to be preferred in all circumstances to reduce environmental burdens from freight traffic. Rail transport only performs much better than road transport if only electric trains are used, relatively heavy goods (kg/m^3) are transported and the trains do not run faster than 70–80 km/h. Even then, the cost-effectiveness of rail investments is extremely poor. It is much more promising and cost-effective to enhance the performance of the modes that cause the most trouble: truck and barge.

Note that reducing NO_x emissions of diesel trains by lowering emission factors of new trains is very probably cost-effective. However, because there are currently no EU regulations for diesel trains, and because rail freight transport is very often international transport, achieving policy progress with respect to diesel trains is more complicated than with respect to trucks.

6. CONCLUSIONS

This chapter has tried to establish that policies aimed at improving the quality of service of bus and rail in order to achieve significant transfer from road traffic are doomed to failure. Evidence shows that rail investments and public transport subsidies lead to a small shift from the road and are not effective policies for reducing the demand for automobile use. Such investments must not be made in the belief that this would bring about more than a marginal transfer of trips currently made by car or truck. The automobile has generated new patterns of activity, the great majority of which cannot be served efficiently by public transport, be it rail or bus.

Because the shift from the automobile to PT is so small, combined with the relatively high share of transfers from environmentally friendly modes and newly generated traffic, the net outcome in environmental terms (energy, emissions) of investing in new PT or rail services is most often negative. Since investments in rail infrastructure are very costly, much more so than in other modes, the cost-effectiveness of these investments appears to be very poor in many cases (CPB et al., 1998).

What should be understood as the message of all this? Is it useless or is it perhaps even wrong to invest in rail or in public transport? This would be the wrong conclusion. It depends on where and for what purposes the investments are planned. It depends on the objectives to be achieved.

In many European countries, most of the infrastructure budget goes to rail investments, and of this, the majority goes to high-speed, long-distance connections. This is a questionable policy, and a questionable way of spending limited public money.

The motivation used by politicians and policy makers to invest in rail, or in public transport in general, is that this investment is needed to curb road congestion and to reduce environmental damage by cars and trucks. This chapter demonstrates that these arguments are questionable, if not invalid overall. Tackling car-induced problems is a weak pro-rail/PT argument.

The merits of public transport (rail or bus) are various (see Hillman, 1993, p. 113):

- it improves accessibility to certain vulnerable areas: city centres;
- it serves spatial policies;
- it offers/guarantees mobility to certain population groups;
- it saves money, for example in the long-distance trips market (recreational and business trips).

In conclusion, sound decision making requires that rail and public transport investments be based on their own merits and not in the first place on their impacts on road traffic.

7. IMPLICATIONS

The purpose of this chapter is to warn against too much optimism about the possible positive impact of rail investments and other rail-related measures on road congestion and environmental impact of transport. Do our findings imply that all rail-related measures are 'bad' measures? In our opinion this is not the case, for several reasons. First, a rail investment project can have a positive result (for example more benefits than costs), regardless of the impact on road congestion and the environment. Second, rail investments can be combined with other measures in a package approach. This package can have a positive result, whereas the rail investments are a necessary part of the package. For example, a package for a region, including rail investments, land-use policies (high densities in the area near stations) and policies to discourage car use (for example road pricing, parking policies, car-free inner cities) may have more benefits than costs. Also between the major cities in the Netherlands the train can compete relatively well with the car. Further improvements in rail, combined with policies to discourage car use, as mentioned before, may result in less congestion, environmental improvements and more benefits than costs. Another example may be possible (future) policies to reduce transit road haulage in Alpine countries. To make such policies acceptable the provision of rail transport as an alternative may be necessary. Third, investments in new stations along existing rail lines may have a much better cost–benefit ratio than building new rail lines (including new stations), especially at places where opportunities to build in high densities are favourable and where rail has a relatively good market position compared to road.

However, we think that very often the assumed positive impacts of rail investments are not realistic, making many rail investments too expensive compared to their benefits.

NOTES

1. Car emission factors in gram/km of some pollutants (e.g. CO_2; SO_2), are higher under congestion than under free-flow conditions. However, NO_x emission factors for cars are lower, due to a lower engine temperature during congestion. Most lorry emission factors are much higher under congestion than under free-flow conditions (Schallaböck and Petersen, 1999).
2. Kinnock uses data for 1990. The increase in congestion levels very likely exceeds the increase in GDP. Therefore the current value (2000) – and assuming the same calculation method – is probably higher. The value of 2 per cent has been used very often since the publication. However, it might be a strong overestimation of the real value – see, for example, Bovy and Salomon (1999).

REFERENCES

Bovy, P.H.L. (2000), 'Traffic flooding the low countries: how the Dutch cope with motorway congestion', Transport Reviews.

Bovy, P.H.L. and I. Salomon (1999), 'The spread of congestion in Europe', in ECMT (ed.), *110th CEMT Round Table*, Paris, pp. 85–153.

Bovy, P.H.L., J. Van Der Waard and A. Baanders (1991), 'Substitution of travel demand between car and public transport: a challenge to policy makers', *Proceedings PTRC Summer Annual Meeting*, Seminar H.

Central Bureau of Statistics (CBS) (1999), *National Travel Survey Yearly Report* (in Dutch) Voorburg (The Netherlands): CBS.
CPB, RIVM, SCP, AVV (1998), *Choose or share: Assessment of investments* (in Dutch), The Hague: Central Planning Bureau.
Centre for Environmental Studies (CE) (1997), *Energy savings in traffic and transportation by spatial planning* (in Dutch). Delft: CE (on behalf of NOVEM).
Commissie Betuweroute (1995), *Report of the Commission Betuweroute*, The Hague: Commission Betuweroute.
ECMT (ed.) (1999), 'Traffic congestion in Europe', *110th CEMT Round Table*, Paris.
Gerondeau, C. (1997), *Transport in Europe*, London: Artech House ITS Series.
Gijsen, A. (2000), *Rail in model. Description and application of the model PRORIN*, Bilthoven/Utrecht: National Institue of Public Health and the Environment/Utrecht University (in Dutch).
Gordon, P. and H. Richardson (1991), 'The commuting paradox', *Journal of American Institute of Planners*, **57**(4), 416–20.
GoudappelCoffeng Consultancy (1997), *Market profile of the jam driver* (in Dutch) Deventer (The Netherlands) (on behalf of Dutch Ministry of Transport).
Hillman, M. (1993), 'Public transport: an unrealistic substitute for most car trips', *Proceedings PTRC Summer Annual Meeting*, Seminar H, pp. 209–19.
Janse, P. et al. (2000), *In search for environmental gains on the rail. Synthesis of investigations of environmental impacts of goods transport by rail* (in Dutch). Delft/Bilthoven: CE/RIVM/TNO-Inro.
Kinnock, N. (1995), *Towards fair and efficient pricing in transport*, Brussels: European Commission.
NEA (1998), *Congestion costs on the main road network 1997*, Rijswijk (The Netherlands): NEA (in Dutch).
Peeters, P.M., P. Rietveld, F. Bruinsma, D. van Vuuren and A.J. Rooijers (1998), *At what time do you think you will be home? A survey on reliability of transport systems and the impact on travel behavior* (in Dutch), The Hague: Projectbureau IVVS.
Rubin, T.A., J.E. Moore II and S. Lee (1999), 'Ten myths about US urban rail systems', *Transport Policy*, **6**, 57–73.
Rutten, B.J.C.M. (1995), 'Does intermodal transport decrease energy consumption and CO_2 emissions in freight transport?', Paper WCTR, Sidney.
Schalläböck, K.O. and R. Petersen (1999), 'Introductory report', in ECMT, *110th CEMT Round Table 110, Traffic congestion in Europe*, Paris: European Conference of Ministers of Transport.
Stembord, H.L. (1991), 'Quality of service on the main road network in The Netherlands', in U. Brannolte (ed.), *Highway capacity and level of service*. Rotterdam: Balkema, pp. 357–65.
TNO–Inro (1998), *RUA Corridor study* (in Dutch) Delft: TNO-Inro.
Transpute (2000), *Research into stabilization of delay times at structural bottlenecks. Testing Gattis' theory on maximum accepted delay in queues* (in Dutch), Gouda (The Netherlands): Transpute (on behalf of Dutch Ministry of Transport).
Van den Brink, R.M.M. and G.P. van Wee (1997), *Energy use and emission per mode*, RIVM-report no. 773002007, Bilthoven: Rijksinstituut voor Volksgezondheid en Milieu.
Van der Waard, J. (1990), *Draft handbook elasticities* (in Dutch), Rotterdam: Rijkswaterstaat, Dienst.

8. Congestion in Europe: Measurements, patterns and policies

Piet H.L. Bovy and Ilan Salomon

1. INTRODUCTION

Congestion has become an inseparable characteristic of many transportation systems. Transportation systems are developed to support public welfare and facilitate economic growth by providing accessibility. Greater mobility is usually associated with greater welfare. However, the evolution of mobility, in both the qualitative and quantitative aspects, has developed to such levels that in many places and times it generates significant negative impacts. These include externalities, such as congestion, environmental pollution of various kinds and safety costs. A congested transport system may fail to deliver sufficient economic benefits, and may have negative ramifications on the competitive position of a region in the European Community. Transportation policy making is thus becoming an art of balancing the desired improvements in mobility and the minimization of its costs to levels deemed acceptable by society.

From the perspective of European policy analysis, the congestion issue is directly related to policy questions such as planning and investment decisions in TERNs (Trans-European Road Networks), financial support to countries and regions in developing international and interregional road links, and the question of fair and efficient pricing of transport in infrastructure and transport use (Kinnock, 1995). These and other policy issues require a much deeper understanding of the congestion phenomenon and of the impacts of congestion.

Congestion is experienced daily not only by many road users, but also by rail and airport travellers, as well as by shippers of freight on these modes. As road congestion is probably the most usual form of congestion experienced daily by literally millions of travellers, the chapter focuses on this type of congestion. Another reason for this focus is that congestion in other modes of transport may call for different solutions. An examination of European statistics highlights the crucial role that road networks fulfil in Europe and the consequent congestion problems encountered on some of these roads. Roads are the most versatile elements of transportation infrastructure, as they serve both passengers and freight, both private and public transport, and they are available to individual operators (of cars) as opposed to the restricted use of rail. Thus, some 85 per cent of all passenger kilometres in Western European countries are made in cars and vans, and high shares of freight movement are on roads (Ministry of Transport, 1996). Roads are the most fundamental element of the transport infrastructure, and consequently, a deterioration of their performance due to congestion is drawing much attention in Europe, as elsewhere.

The private and external costs of congestion are generally considered quite substantial. The spread of congestion across Europe, as well as many other parts of the world, is worrying and consequently it has attracted a considerable effort on the part of policy makers and researchers trying to identify policies that can mitigate its effects and reduce its costs. But there are some doubts regarding the effectiveness of many such policies and even regarding the rationale of congestion-related policies.

This chapter addresses the congestion issue from a Western European perspective, by focusing on a number of key questions:

- What are the current patterns and trends of congestion?
- Should congestion be eliminated altogether?
- What can and should be done to mitigate undesired levels of congestion?

Section 3 focuses on the background factors. It describes the various external and internal dynamics which produce congestion. Section 4 dwells on the measurement issues and describes congestion patterns in Europe, relying on a variety of comparative data.

In recent years, congestion seems to be expanding in its temporal and spatial distribution. This raises a number of questions which will be addressed in the following sections:

- What would happen if it were left without any policy response?
- What can be done to change the trends?
- What are the relevant policy responses?

To respond to these questions, we discuss congestion from a behavioural perspective in section 5. The main point of this individual traveller's perspective is to demonstrate that the problem seems very different from that perspective as compared with the system-wide perspective. The implications of this perceptual gap are then discussed in section 6, which focuses on coping with congestion. It opens with a discussion of whether or not it is desirable to have some level of congestion. It then presents a review of various approaches to address the congestion problem where and when it exceeds a desired level.

Section 7 presents the main conclusions, followed in section 8 by some brief recommendations. The chapter is a much abridged version of a report written on behalf of ECMT (see Bovy and Salomon, 1999). For details of this chapter, see the ECMT report.

2. ARE THERE EUROPEAN CONGESTION PATTERNS OR PROBLEMS?

Congestion seems to be increasing in many parts of the world, from the North American megalopoles to the Western European conurbations and to the rapidly growing metropolitan areas of Southeast Asia. The focus on Europe implies that there are some unique attributes of European congestion which are absent in or different from those in other parts of the world.

While congestion may be viewed as a form of queue, where passengers are waiting to traverse a particular link or node where demand temporarily exceeds supply, its underlying causes differ across locations and times. The focus on Europe is motivated by the fact that travel patterns and trends in Europe differ from those in North America, Japan or other developing metropolitan areas across the world. One of the explanations for such differences lies in the time factor. The timing of the introduction of various technologies and social trends across different parts of the world may explain why present conditions and trends vary in different regions.

As will be shown, the private automobile, which lies at the basis of the congestion problem, gained its popularity in Europe at a different time and against a different spatial, economic and social environment compared to other parts of the world and, consequently, congestion patterns have evolved in a different way. The difference in the phasing of the growth in congestion also implies, as will be shown below, that the range of relevant policy measures appropriate for implementation in Europe probably differs from those relevant for America or Asia. Two other important conditions for congestion development differ significantly between the continents, namely the spatial settings and availability of travel options and transport alternatives to the car.

But one should note that a European focus must be qualified as well. Within Europe, there is a wide amount of variation. In particular, it seems that congestion is not a continent-wide phenomenon and is not likely to become one. A continent is not the proper unit of analysis for such a study. Congestion is clearly a regional phenomenon, concentrated in the densely populated areas of northwestern Europe, as much as in highly urbanized areas in the rest of the world. Consequently, it is also quite irrelevant to compare cross-national statistics on congestion, as these comparisons conceal more than they illustrate the differences. Regions seem to be the appropriate unit of analysis. Thus, in this chapter we provide data from various regions in Europe, rather than national travel patterns. But, for background information, it is worth examining some of the basic differences between Europe and other developed economies (see Salomon et al., 1993; Pucher and Lefevre, 1996).

3. CAUSES OF CONGESTION

External Factors

Traffic congestion is the result of a multitude of factors. The importance of each factor varies from one place to another and across time. Broadly defined, the causes can be attributed to demand and to supply factors, but these are, of course, at some point, interrelated. In the following sections two complementary explanations for the evolution of congestion are presented. The first focuses on the external forces that increase the car dependence of the population in developed countries. Second, the internal dynamics of congestion will be described, to demonstrate the processes by which changes on a network occur in the presence of congestion.

Figure 8.1 presents a flow chart of the main effects that are at play on the demand and supply sides. There are, as will be noted, many additional effects and feedback mechanisms, but for the purpose of organizing the description of the factors only the main effects have been drawn.

Socio-demographic factors

The driving population is growing and one of the consequences is a growing demand for travel by car. The growth in the driving population is the result of a number of background trends. First, there is growth in the population. But more important is the fact that the *household population* is growing faster than the general population, in particular in the more urban segments of the European societies. This is, in part, attributed to the growth in smaller households (of single persons, single parents and smaller numbers of children in the households). As households are independent units of consumption and production, more households imply more maintenance trips and a greater demand for automobiles.

The *ageing* of the population, as a result of prolonged longevity, involves yet another contribution to the size of the driving population. The elderly of today and increasingly in the future are more likely to own a driver's licence than in the past.

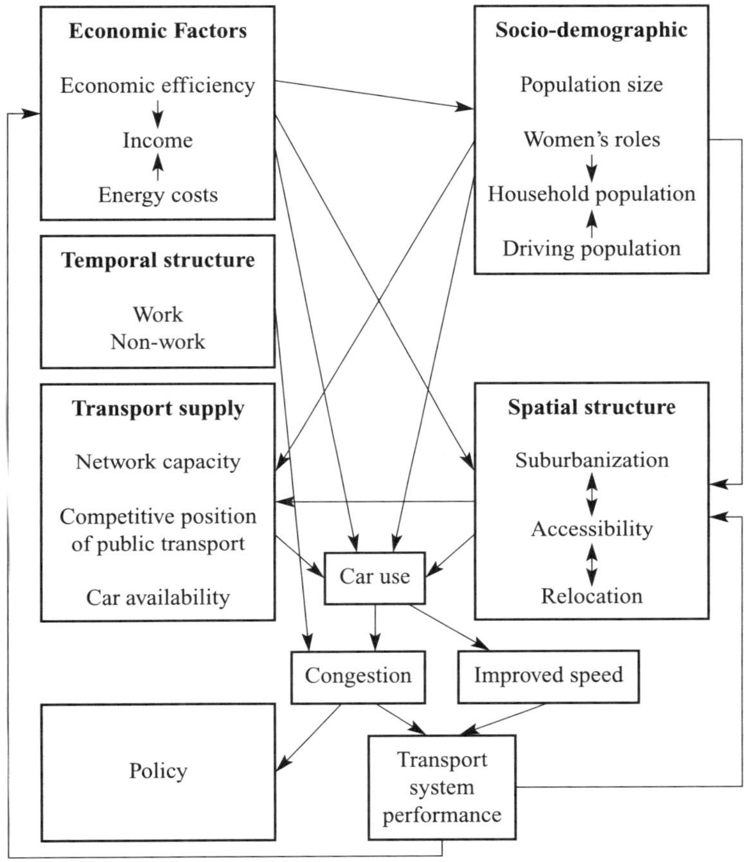

Figure 8.1 The main external factors affecting congestion

Economic factors

Growing income has brought about a general rise in the standards of living and the automobile has become an integral part of these new standards. Coupled with the relatively low costs of automobiles and their operation, the availability of the car for a growing number of activities has become the norm. Generally, energy prices in Europe are significantly higher than in North America, but still, auto usage is relatively cheap. Growing income effects, among other things, changes in residential location, as it facilitates the acquisition of private houses in suburban locations.

The growing *car population* is a major contributing factor to congestion. Cars are produced in response to demand which is growing steadily as people realize the convenience of private transportation and the growing utility of using a car, given its costs and its advantages. The automobile industry in many European countries is an important element in the national economy and is supported directly or indirectly by social structures. This in itself is a contributing factor to the growing popularity of the private automobile.

The costs of cars and their operation are taxed in all countries, but the structure of the various relevant taxes differs in the signals it generates with regard to auto usage patterns. In most cases, gasoline taxes (and parking taxes, as opposed to rates) are the only usage-based taxes. These seem to be relatively weak and have no bearing on congestion, as they do not reflect spatial and temporal variations. Only a few countries use road pricing as an instrument to influence the use of the road infrastructure (for example France, Italy, Norway).

Spatial structure

From a historical perspective, the relationship between suburbanization and congestion can be divided into two periods. Initially, with the suburbanization of residences, congestion was primarily evident on radial links of the network. Later, with the growing suburbanization of employment and commerce, congestion started to become a problem of suburban regions, on both radial and circumferential links of the network.

These changes also have implications for the temporal and spatial distribution of flows and consequently for the likelihood of experiencing some levels of congestion. When trips were primarily centre-oriented (and work schedules were quite fixed), flows followed a pattern of an inward-moving wave, with congestion becoming more acute closer to the centre. The current patterns of congestion are more complex, both in terms of location and timing.

Activity-related factors

The demand for travel, except in some relatively rare situations, is derived from the demand for activities performed at the trip ends, which in turn are determined by the *lifestyles* individuals wish to exercise, the spatial distribution of opportunities (*land-use patterns*) and the *temporal structure* which prevails in a given society.

The *temporal structure* lies at the heart of the congestion problem. The prevailing temporal structure, which is very much a culturally dependent factor, explains much of the activity patterns, especially with regard to daily work schedules, and daily and weekly shopping patterns.

The distribution of trips over time in each of the daily peaks depends on the range of official work start times, the range of flexibility permitted in the workplace and the level of congestion experienced by commuters.

Activity behaviour refers to work and non-work personal scheduling. Changes in work scheduling due to congestion include both daily working times and working days, and chaining of daily activities. The same may be applied to non-work out-of-home activities where people change their schedules. The main motive for work rescheduling will be decreasing reliability and travel time duration.

A Systems Dynamics View of Congestion Causes

The road congestion problem as it developed in the last decades is a typical example of a self-reinforcing process with short- and long-term feedback loops stimulating car use. Figure 8.2 describes, in a simplified and condensed systems dynamics flow chart, the essentials of the mutual influences (on the level of individual households and firms) that endogenous factors in the economic and transportation system exert on each other.

For the sake of clarity, the exogenous factors (described in Figure 8.1 above) that are simultaneously at work are omitted from Figure 8.2, despite their clear influence on the levels of car use and traffic congestion.

Car availability and car use (in terms of distance travelled) are the basic engines of the process, fed by available incomes. Car use leads to higher door-to-door speeds, which enable individuals and firms to cover a much larger range for their activities with much higher utilities (including gaining a higher income) achievable within the same travel-time budget. An important part of these higher utilities is, for example, lower land and housing costs, leading in turn to spatially dispersed settlement patterns. These increased travel distances, combined with increased trip numbers due to demographic growth, have resulted in growing demand for road space and have required extensions of the road network. Through the establishment of extensive motorway networks in Europe, medium- and long-distance door-to-door travel times have been shortened dramatically, thus speeding up the described spatial transitions. At the same time, the improved roadway system has contributed significantly to higher economic performance and therefore higher income and lower car costs. In fact, car costs per travelled kilometre (for the same level of driving quality) are continuously decreasing. This is another strong force for further increases in car ownership and use. A third feedback loop is the diminishing competitiveness of the alternatives to the car (walk, bike, public transport) mainly due to spatial dispersion and larger distances.

These circular processes have to be taken into account in the development of policies aimed at controlling road traffic congestion. A crucial factor is door-to-door speed. Further improvements in speeds lead to a further proliferation of the system. A critical element in any congestion

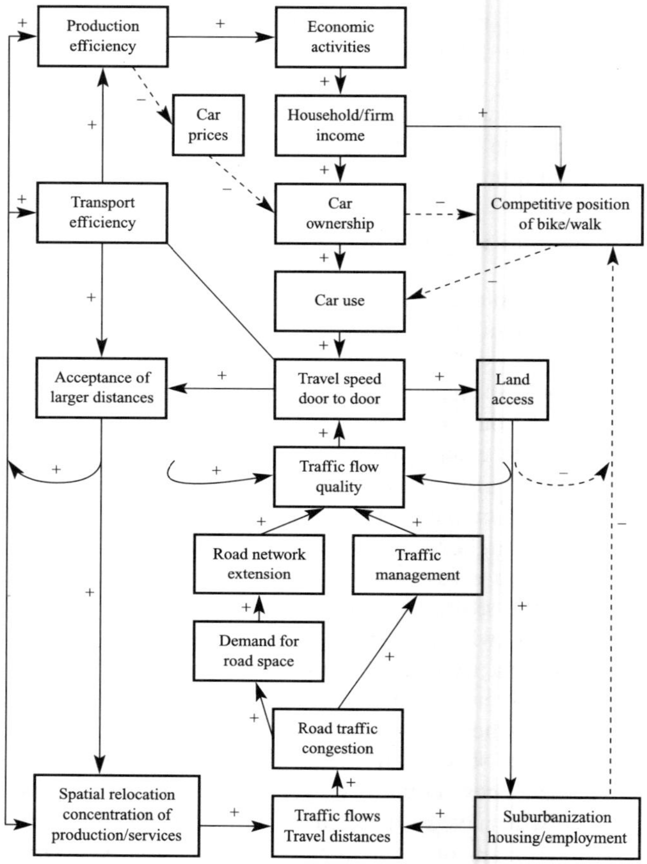

Figure 8.2 System dynamics model of factors contributing to congestion

policy (infrastructure extension, congestion pricing, and so on) is therefore to contain travel speeds within economically tolerable limits. Congestion policy measures should, therefore, be directed predominantly at offering the required capacity at an economically sound level of service, but without further increasing travel speeds.

4. THE NATURE AND EXTENT OF EUROPEAN CONGESTION

Perspectives

The colloquial explanation for congestion is that insufficient road capacity has been provided. This view is based on a widely accepted notion that road space is a free public good that needs to be supplied by the authorities to accommodate any level of demand. However, alternative views can be suggested. First, this reasoning can be turned around and it can be suggested that congestion is a result of excess use of vehicles, rather than insufficient supply of road capacity.

But not all congestion is a result of insufficient supply. Some congestion is not recurrent and is a result of particular temporary conditions like accidents, severe weather conditions and road maintenance work. Recurrent congestion is caused by a structural lack of capacity (or equivalently excess demand), whereas non-recurring congestion stems from incidental lack of capacity or excess demand. It is important to note that even in cases of recurrent congestion the travel characteristics can change from day to day: it is uncertain where queue building starts, when it starts, when it ends, how large waiting-time loss will be, and so on. So, apart from the existence and travel-time losses of queues, the unreliability of queue location and duration is a major aspect of congestion which greatly affects travellers' behaviour.

Congestion is a double-faced phenomenon: on the one hand it may be viewed as an attribute of the network; it can be described, for example, by the number and length of queues that have occurred in the network, or by their duration. Similarly, the level of congestion may be described by the length of the network which was affected by queues. This network-related view is most common in official statistics and in the public debate.

On the other hand, congestion is also an attribute of a trip. This perspective on congestion is of interest because it entails some important attributes which influence the traveller's behaviour. Such trip-related characteristics are, for example, whether a trip encounters congestion (percentage of trips experiencing congestion), amount of time or distance travelled under congested conditions, and share of delay time to total trip time. Unfortunately, empirical data on trip-related congestion variables are very rare.

The extent of congestion can be demonstrated by some statistics of the Dutch Randstad area (Ministry of Transport, 1997). On an average working day in 1996 about fifty queues of minimum length of 2 km build up, mainly at the fringes of the four big cities. As indicated above, congestion is highly variable, so the number, location and times of queues change from day to day. Bridges and tunnels across major waterways are well-known queue locations.

Also discontinuities in the freeway network (entries, exits, weaving sections, lane-number alterations) are favourite queuing places. The typical location of recurrent queues around the bigger cities stems in a large part from changes in the spatial orientation of travel demand in the last 20 years. The changes include developments such as reverse commuting and criss-cross travel demand between suburbs.

Concerning the timing of congestion, 80 per cent of the queues in the Randstad occur during the peaks, of which 45 per cent in the morning peak (from 7.00 to 9.00 a.m.) and 35 per cent in the afternoon peak (from 4.00 to 6.00 p.m.). From the average of 40 daily queues about ten queues take place during off-peak times. These are mainly queues caused by incidents.

Measures of Congestion

Unit of analysis

Congestion is an important notion in transport decision making. It is a relevant quantity in network design, facility dimensioning, and pricing strategy. It is therefore striking that policy makers are still struggling for a clearly defined, unambiguously measurable indicator of congestion. There are only very few countries (for example the Netherlands) with reliable congestion statistics (see for example Ministry of Transport, 1996 and NEA, 1996).

A brief discussion of the problem of measuring the amount of congestion in transportation systems is presented below. It is intended to contribute to the ongoing European efforts to produce a standardized approach (see Annex 3 in ECMT, 1995, for a proposed method for harmonizing measurement of road congestion).

It is fundamental to distinguish two classes of congestion measures. One class is related to flow conditions on the network, and the other to parameters of travel conditions between origins and destinations. In the first class of measures, the unit of observation is a link in the network, and we may look at volumes, speeds, or link traversal times to derive values for the level of link congestion and, possibly, the related costs for that link (ECMT, 1995). This may be done using traffic assignment models (Transroute et al., 1992), or on the basis of observed link volumes (for example, ECMT, 1995, Annex 3). Adding up the link values results in a network-wide figure. This approach can also be applied to specific network categories, such as non-urban motorways, urban radial arteries, and the like. Most of the data on traffic congestion belong to this class of network congestion measures (see, for example, Bukold, 1997, and ECMT, 1995). These network-related congestion measures are indicators of the quality of the network performance.

The mirror image of the network performance attributes are measures of congestion as they apply to the individual's travel behaviour. These measures refer to trip characteristics of travellers, from their respective origins to their destinations. In this type of measure, the trip or tour is the unit of observation. The quality of a transport system can thus be described by the effects of congestion on travel conditions, on travel choices, or on how congestion pricing might affect travel behaviour. Such measures refer to the congestion experienced by users and include the number of times a driver encounters a queue, the duration of waiting, and the total excess trip time due to congestion.

The rationale behind taking the trip as a unit of observation is that trip-making behaviour such as route, mode and departure-time choices are based on the characteristics of the entire trip. In this respect it would be even more relevant to address a round trip (origin–destination–origin) as the relevant unit of analysis. Information on trip-related congestion and its derived costs is of utmost importance for policy making, but is unfortunately rarely available. These measures, similar to the classification of network measures, can be classified by the type of network upon which various segments of the trip are made.

Presumably, correct accounting of congestion losses by both classes of measures should lead to the same levels. However, the two classes of measures differ in the information they convey. Some examples will highlight the differential sensitivity:

(a) In certain bottlenecks of the road network the total congestion duration may be long and last for hours. The bottleneck serves high traffic volumes, so that the individual waiting times at the bottleneck are short, on the magnitude of a few minutes. This is a negligible amount from the individual's perspective and does not lead to adaptation of alternative travel choices.
(b) The additional time spent in bottlenecks may be fully or partly compensated by quick and easy progress because of high speeds for the other segments of the trip, such that total trip time remains within individually acceptable values.
(c) Network congestion measures only measure revealed congestion as such. They do not measure the impact of congestion on users who avoid the congestion by adapting their behaviour, such as drivers who take a detour route or an earlier departure time with a longer trip time, but without congestion.

Thus network congestion figures, on the one hand, may provide a picture which is too negative, as they are based on the accumulation of a large number of small, behaviourally irrelevant queuing times. On the other hand, they fail to account for the impacts of congestion on suppressing demand for travel. Therefore, trip-based congestion measures are to be preferred.

The distinction between network and trip-based congestion indicators can explain the paradox identified by Gordon and Richardson (1991) which states that while aggregate congestion figures appear to increase steadily from year to year, average travel times, speeds and congestion losses per trip remain more or less constant. Despite the increasing congestion on motorways (in aggregate figures) such roads do not seem to lose their attractiveness, as is evident from growing volumes. The explanation is that congestion experienced by individuals does not increase significantly.

In addition, it is hypothesized that drivers are willing to accept a certain waiting time at bottlenecks. It seems that up to 10 to 15 minutes of queuing is acceptable, and only beyond this level do drivers engage in adapting their behaviour to alternative travel patterns. This hypothesis is derived from the observation (in the Netherlands) that at a number of classic bottlenecks the maximum queue length has remained stable for many years (Westland, 1997).

Measurement of congestion: a critique

Having discussed the unit-of-analysis problem in measuring congestion, the question of the correct way of measuring congestion losses remains in either case.

Congestion may be defined as a state of traffic flow on a transportation facility characterized by high densities and low speeds, relative to some chosen reference state (with low densities and high speeds). It should be stressed that high flows are not typical for congestion; in many instances the congestion state results in low flows and low speeds. Flow levels alone are thus not a useful indicator for congestion. What the reference state (zero congestion) should be depends, among other things, on the purpose of decision making (infrastructure decisions, traffic management decisions, congestion pricing decisions, and so on).

Both from a public-policy-making perspective and from an individual travel decision-making point of view, the congestion burden should be translated into costs. For the sake of clarity and comprehensive accounting, it is important to distinguish between congestion costs of the following four groups of transportation system users who are affected by (rising) congestion:

(a) higher travel costs for road users who use bottlenecks and experience congestion (including the costs of unreliability due to variability in congestion);
(b) higher travel costs for road users who avoid congestion, for example by changing route or departure time (suppressed bottleneck demand);
(c) higher travel costs incurred by other users of the transportation system, due to demand shifts caused by congestion, for example shifts of road users to public transport (suppressed road traffic demand);

(d) reduced benefits due to a change in activity and therefore derived travel pattern (suppressed travel demand).

Cases (a) to (c) imply a lower consumer surplus, given the same level of activity is maintained and the same benefits are accrued. In all four cases, both private costs (borne by the congestion-causing road user) and external costs (borne by others) may be involved.

We may now define congestion costs as the additional costs caused by the existence of congestion, relative to some adequately chosen reference state.

In looking at the European congestion costs, published in official European documents (for example, Green Paper, OECD, CEMT; see DHV/Colquhoun, 1991), a disturbing variety of cost figures emerges. This results, in part, from the fact that a variety of methods is applied, but also from the confusion of the private, external and social costs of congestion. In Kinnock's well-known Green Paper (Kinnock, 1995), the external congestion costs in Europe, that is, that part of the costs not borne by those who cause the traffic congestion, is stated to be about 2 per cent of GDP. This cost level was quoted from an OECD survey (Quinet, 1994) which, however, stated that social costs of congestion totalled 2 per cent of GDP! In addition, the OECD survey restricted its calculation of social costs to the additional travel costs of travellers who experienced congestion. Thus these costs include the private costs (which are matched by private benefits) and only part of the external costs. Generally, the calculated additional costs in the survey referred to the free-flow reference situation, which is a questionable approach, and in most cases was calculated using static network assignment modelling which is a clearly deficient procedure to estimate congestion costs.

One may confidently state that the often-quoted 2 per cent of GDP congestion cost figure lacks a clear empirical and methodological foundation, and is not more than a initial rough guess unsuited for serious policy making. Calculations by Gerondeau (1997) show that a figure of 0.3 per cent might be equally plausible (this equals, for example, the congestion cost level calculated for Dutch motorway traffic based on an extensive congestion monitoring system).

Almost all figures found on congestion costs in Europe are derived from the 'additional time spent' travelling, relative to a chosen reference situation. Using appropriate value-of-time estimates for travellers and goods, and vehicle fuel consumption, this extra travel time then is transferred into a monetary cost figure. A few comments can be made with respect to the choice of the adequate reference situation (see also Gerondeau, 1997):

(a) Some congestion estimates have used as a reference a collection of ideal door-to-door trips, based on distance calculations determined by a fixed detour factor relative to the airline distance (say 1.2), and with certain ideal travel speeds (say 100 km/h outside and 50 km within urban areas). In one of the European infrastructure studies (DHV/Colquhoun, 1991) zone-to-zone minimum speeds were adopted (90 km/h for cars, 80 km/h for trucks) to determine the level of inadequate performance of the network. When comparing the actual travel conditions with such a reference, one is in fact calculating the costs of an inadequate road network instead of the costs of congestion.

(b) The most frequently applied approach is based on the existing road network as given and considers the 'empty' network as the reference point. This means that the actual traffic pattern is compared to trips on shortest routes at maximum speeds, even in peak periods. Clearly, such an empty network is an unrealistic yardstick. And more importantly, a network satisfying such conditions would look quite different and would be very inefficient as well. Nevertheless, the studies using this approach formed the basis for the OECD survey on congestion costs (Kinnock, 1995).

(c) A few studies have used the assumption that low speeds necessarily imply that congestion exists, or that high volume/capacity ratios are unambiguous indicators of congestion (Transroute et al., 1992). The level-of-service concept is often used to define the reference conditions (for example Bukold, 1997). This is clearly a more reasonable approach because one can choose the conditions in which the network optimally fulfils its transportation function. In the Netherlands, economic calculations have been performed to derive a social cost optimum for traffic flow (Stembord, 1991). It appeared that a 2 per cent congestion probability (which means that on a yearly basis 2 per cent of daily traffic of a road section will be experiencing a queue) is the optimum level of congestion. This level serves as a reference for quantifying the costs of additional travel time.

Measures of congestion: a proposal for improvement

In summary, estimations of the economic costs of congestion exist in many European countries and on the continent-wide level, but their outcomes are so different and are based on such widely diverging assumptions and methods that their credibility is very poor. Policy making in Europe with respect to congestion needs to be based on valid and comparable facts, and on measured and estimated congestion characteristics of the infrastructures and of the trips. Ongoing work on improving congestion measuring methods (see for example WP5, 1994) should be forcefully continued. Our inventory of congestion figures clarified that a much more rigorous and systematic analysis of congestion costs is needed in European countries, exhibiting, among other things:

(a) a clear distinction between private and external costs;
(b) a clear distinction between road users and non-users;
(c) a clear distinction between travel costs and other congestion costs;
(d) a clear definition of the reference situation, preferably based on a standardized, economically optimal network design;
(e) a standardized and valid calculation procedure of the cost elements;
(f) congestion figures for both network elements and for trips;
(g) flow and speed data with reference to hours.

Network Congestion Patterns in Europe

Comparative studies

A few studies have tried to estimate congestion characteristics of European roads to obtain an overall picture of the state of the network, in terms of spread, scope and costs of road traffic congestion in Europe as a whole, or at the country level.

One of these studies (Transroute et al., 1992) shows that problems of congestion are being experienced on more than 5000 kilometres (including 3800 kilometres of motorway) of the 54 000 kilometres of roads of international importance within the European Community (half of which consists of motorways). This means that nearly 10 per cent of this high-level network is affected. As will be demonstrated below, this figure of 10 per cent is an average value with large variations between countries, and even more between regions, and with strong spatial concentration.

In an ECMT survey, member countries were asked (1991/92) to report traffic congestion on their main roads (ECMT, 1993; see also ECMT, 1995). Figure 8.3 depicts the survey results in a scatterplot of congestion points. Unfortunately, the results from the different countries are not comparable. The quality and quantity of responses were strongly correlated with the level of traffic density in each country. In addition, there was a large variation in thresholds used to define congestion. Despite this, there is a clear spatial pattern in the congested spots, such as around high-density conurbations (London, Paris, the Randstad, the Ruhr area, Athens, and so on). The Scandinavian countries clearly appear to suffer least from road congestion.

From this and other studies (Bukold, 1997) it can be clearly seen that the contribution of international traffic to congestion is very limited. There are virtually no cross-border links that suffer from congestion, which is not surprising given the relatively small international flows in absolute and relative terms.

None of the studies reviewed shows the specific international dimensions of congestion such as the contribution of international traffic. The same holds for the specific contribution of road freight transport, with the exception of

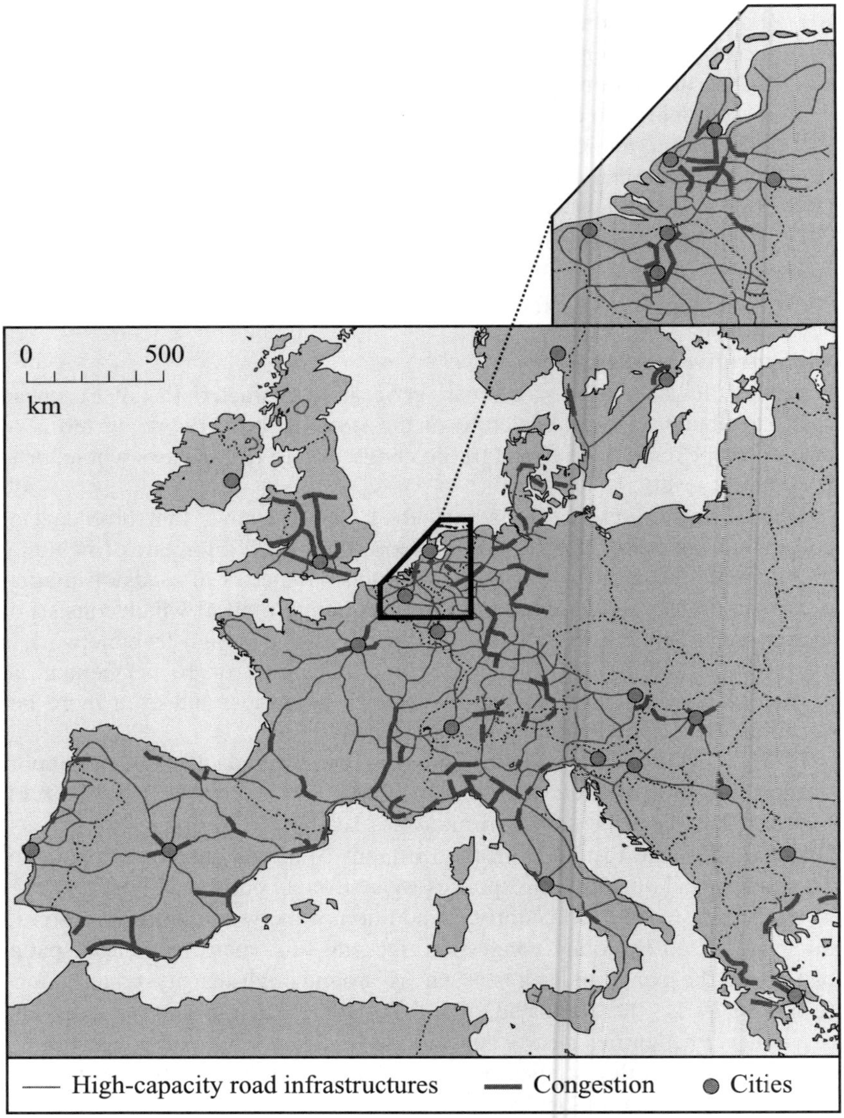

Source: CEMT (1995).

Figure 8.3 Map of major congested links in the European road network

the DHV/Colquhoun study, which showed that from the total yearly cost of 350 million ECU due to inadequate level of service to road traffic, only 50 million can be attributed to freight traffic (1990).

Perceived quality of road infrastructure

An international comparison of European road traffic congestion suffers from lack of readily available and recent data. Comparable data on congestion based on a sound measurement methodology do not exist (with one exception discussed below). Nevertheless, some indicators of congestion in different countries can give rise to several hypotheses on differences in causes of congestion.

Road traffic congestion is related to the level of road infrastructure supply relative to the demand for trips. Table 8.1 provides statistics on network supply and congestion for some developed countries.

The data in Table 8.1 demonstrate that even in countries with a similar level of economic development, the supply of road infrastructure per capita differs by more than 100 per cent. The level of congestion, as measured by a relatively objective indicator (column 3), clearly appears to be related to the

Table 8.1 Road quality rated by international business panel

	Road network (km/100 inh.) 1993 (1)	Motorways (km/mn inh.) 1993 (2)	Congestion (% of links) 1993* (3)	Perceived road quality 1995** (4)
USA***	14.5	331	–	9.0
Japan	6.2	37 (1987)	–	6.2
UK	6.2	56	24.1	5.9
Germany	7.6	136	7.9	8.3
France	15.8	129	4.5	8.5
Netherlands	6.1	141	14.8	5.9
Belgium	12.9	169	5.9	8.3
Denmark	13.7	127	0.0	9.1

Notes:

* % of motorway links with more than one hour of congestion per day (Bukold, 1997).

** Scale of 1 (low) to 10 (high) based on assessment of international business panel (IMD, 1996).

*** Data for the USA include the interstate system plus the urban freeways, for 1994.

Source: Ministry of Transport (1996).

level of supply (columns 1 and 2). The relationship is corroborated by a perceived quality indicator reported by an international business panel (column 4).

Despite their significantly larger capacity, congestion problems are at present predominant on (urban and suburban) motorways.

A comparison on a national scale certainly does not realistically reflect the typical characteristics of traffic congestion. Some more location-specific comparative analyses will therefore be added.

European congestion: some comparisons

Perhaps the first truly comparative study on the distribution of road congestion in Europe is the ECIS investigation on *Bottlenecks in European Infrastructure* (Bukold, 1997). The study provides a comparable description of the current conditions (1993/94) in the European major road network, consisting of some 13 000 links and including all motorways. To this end, all national networks have been examined using a standard set of performance indicators, namely the commonly used level-of-service (LOS) measure, as defined by the US standards. A congested bottleneck is characterized by the LOS categories E (low speeds, unstable flow) or F (stop-and-go congestion).

Not surprisingly, the links with the highest traffic flows, exceeding 50 000 vehicles per day, are found in the highly populated conurbations. These include London, the UK's north–south corridors, the Rhine corridor (Randstad, Ruhr, Rhine–Main), Paris and the Rhone valley, northern Italy, and also the Madrid and Barcelona regions. According to the conventional LOS criterion, most bottlenecks are found in the UK, Spain and the Rhine corridor, and to a lesser extent also in Austria, Poland and the Czech Republic. By contrast, France's road capacity appears to be sufficient, with the exception of a few urban areas.

A more detailed insight into bottleneck situations is provided by observing the duration of congestion on links in the road network. The number of congested hours provides a more accurate picture of the actual magnitude of the congestion problem in Europe. It appears that severe bottlenecks (links with more than three congested hours per day, on average) are quite limited in number and mainly occur very close to major cities (see Table 8.2).

According to the ECIS study, European countries show striking differences regarding the number and proportion of congested links. Exceptionally high proportions of congested links are found in Spain and the UK. Also the Netherlands and Italy have comparatively high shares of links with severe bottlenecks in their road networks. By contrast, congested bottlenecks hardly exist in Scandinavia. With a few exceptions, the European road bottleneck problem is mainly an urban problem rather than a problem for long-distance connections or cross-border links.

Table 8.2 Percentage of main network links exhibiting a congestion duration of a certain number of hours

	Duration of daily bottlenecks				
Country	0 hours	1 hour	2 hours	3 hours	> 3 hours
Austria	95.5	0	0	3.0	1.5
Belgium	94.1	2.3	0.9	0.9	1.8
Denmark	100.0	0	0	0	0
Finland	100.0	0	0	0	0
France	95.5	0	0.5	0.5	3.6
Germany	92.1	0.6	0.8	1.2	5.3
Greece	98.8	0	1.3	0	0
Ireland	86.2	3.5	0	3.5	6.9
Italy	90.6	0	0.8	2.4	6.3
Luxembourg	100.0	0	0	0	0
Netherlands	85.2	3.8	2.8	3.1	5.2
Portugal	94.9	0	0	0	5.1
Spain	81.1	0.9	1.8	0.9	15.3
Sweden	100.0	0	0	0	0
Switzerland	93.6	0	0	0	6.4
United Kingdom	75.9	3.7	6.5	2.8	11.1

Source: Bukold (1997).

The unique ECIS study allows the drawing of some conclusions:

(a) Congestion in Europe is mainly within and close to urban areas. Improvements should be focused on urban infrastructure (including, for example, urban light rail, regional heavy rail, ring roads, tunnels, local by-passes) for single major cities and for conurbations such as the Randstad and the Ruhr area.

(b) The national situations differ widely. Though almost all countries have bottlenecks (except Scandinavia), only few suffer from heavy congestion. The background factors which may account for these differences are:

   (i) Rapidly growing transport demand as a result of economic development or population growth (for example Spain, Poland);
   (ii) persistent underinvestment (for example the UK); and
   (iii) Environmental constraints or problems of physical limitations (for example the Netherlands and Germany).

(c) Most bottlenecks and heavily congested roads coincide with areas of high population density. Because of apparently severe spatial and environmental restrictions new roads can be a solution only to a limited extent. Packages of road pricing, investments to divert through traffic, and public transport (bus and rail) improvements are key instruments in such cases.

A comparison of three conurbations' road networks

In order to gain insight into the underlying factors of motorway congestion in the Randstad, the Dutch Ministry of Transport commissioned an international comparative study of three similar regions (Hilbers and Verroen, 1996; Hilbers et al., 1997). In those studies, the patterns of use of the main road networks in the Randstad area, the Ruhr area and the Antwerp–Brussels–Ghent region were compared and analysed using background factors such as network supply, spatial conditions, mobility patterns and socio-economic variables.

The regions belong to the high-density conurbations of Europe and are similar in size and structure. Interestingly, however, the level of motorway congestion (expressed as a percentage of the network with more than three hours of congestion per day on average, see Table 8.3) strongly differs. According to the ECIS study (Bukold, 1997), the Randstad network shows by far the highest level, with the Ruhr area only half that of the Randstad level, whereas the Flemish triangle has a negligible number of links with this level of congestion hours.

In line with the ECIS congestion findings, motorway use in the Randstad is considerably higher than in the two other regions: a 25 per cent higher vehicle density on a per-lane basis. A first important explanatory factor is the supply of road infrastructure in relation to the number of inhabitants and the size of the regions. Whereas the motorway supply in the Randstad is more or less similar to the other regions, the big difference is in the supply of the underlying network, which is much less cohesive and dense.

In addition, the Randstad motorway network is characterized by a higher level of accessibility in terms of entry/exit points; also its ring roads (around Amsterdam, Rotterdam, Utrecht) are much closer to the built-up areas than in the other regions. Consequently, in the Randstad more inhabitants have easy access to the motorway and at the same time more inhabitants have to share the same scarce roadway space. These factors explain the relatively high pressure on the Randstad motorways, which are characterized by a higher share of short trips.

A second explanatory factor relates to the level of mobility of the inhabitants in the respective regions. Whereas the Randstad and the Flanders triangle show very similar levels of daily kilometres travelled per capita, the residents of the Ruhr area travel much less (22.5 km). This is partly due to the spatial

Table 8.3 Key roadway and mobility parameters for the Randstad, the Ruhr area and Flanders

	Randstad	Ruhr area	Flanders area
Share (%) of network links with:			
< 1 hour of congestion	85.0	90.0	95.0
> 3 hours of congestion	5.2	2.5	1.8
Daily volume per lane (veh./day)			
All main roads	10 000	8 100	8 000
Motorways	16 800	13 200	13 600
Daily car/km per inhabitant			
On motorways	8.1	6.9	7.7
On other main roads	1.9	2.6	4.3
Network density (km/1000km^2)			
Motorways	115	120	80
Other main roads	105	185	180
Road capacity per capita (lane-km/mn inh.)			
Motorways	480	523	571
Other main roads	320	460	690
Personal mobility (km/capita)			
All modes, all purposes	32.5	22.5	34.4

Sources: Bukold (1977) and Hilbers et al. (1997).

distribution of activities, such as a greater concentration of employment in the city centres in the Ruhr area, and also stronger local orientation of activities for the Ruhr residents. In the Ruhr area, housing demand is largely satisfied by the local market, thus facilitating a much more spatially limited activity pattern. Conversely, the spatial distribution in the Randstad area is much more dispersed, with wider separation between residential areas and employment.

It is interesting to examine the planned infrastructure investments in the three regions (Hendriks et al., 1997). In all three regions, most investments are in rail infrastructure for public transport. In the Randstad and Ruhr areas, this share is about 55 per cent, whereas in Flanders it is nearly 80 per cent. It is important to note that a large part of these rail investments are an improvement of interregional accessibility, especially as part of the European High Speed Rail Network. With respect to road infrastructure it is notable that almost no new links will be built in the coming decades. Instead, most of the investments are spent in extending existing links to 2 × 3 or 2 × 4 lanes. One may safely suggest that the investments in rail, oriented to the service of long-distance travel, will not contribute to congestion relief, whereas the road investments will.

It can be concluded that the high levels of congestion around major urbanized areas are the result of two simultaneous factors, both associated with high population density:

- less space available for road infrastructure; and
- large demand densities (more users per unit of road space).

Individually Experienced Congestion in Europe

Using the scarce readily available data on European congestion, a few observations can be made. In most European countries, the current share of motorway links with less than one hour daily congestion is about 10 per cent or less (Table 8.2). Is that an alarming situation? In official national and EC documents and in the media, congestion losses are totalled to hundreds of daily recurrent queues, millions of hours daily lost in queues, and billions of ECU yearly wasted in road congestion. There is no doubt that congestion is one of the most often cited problems of transportation systems. It is often suggested that European networks are close to collapsing. Is the situation really that bad?

It may be useful to reframe the question. For example, instead of providing huge (and impressive) numbers of hours lost, the question may be posed as: how well does the network fulfil its transportation functions? How many travellers experience congestion regularly, and for how much time?

Before trying to answer this, it is useful to place the network-wide total figures in perspective. In the Netherlands and the UK, the Western European countries suffering most from congestion, total excess time due to motorway congestion is about 2 per cent of total time spent travelling by cars. The related congestion costs turn out to be about 0.25 per cent of GDP. Admittedly, this burden has to be carried by a relatively small part of the travelling population because of the strong concentration of congestion in time and space.

Studies carried out on behalf of the French Road Federation corroborated findings reported in many other national sources that the Europeans who travel daily by car to work (that is 80 per cent of those who use motorized modes) need on average about 20 minutes to get to work (Gerondeau,1997). Not more than 10 per cent of them took more than 30 minutes to get to work. Considering that most of these trips take place during peak hours, this hardly suggests a high level of overall road network congestion. Only 10 per cent of these car commuters reported that they usually encounter traffic jams on their way to work; thus 90 per cent do not encounter congestion, even during rush

Table 8.4 Share of car commuters who experience congestion on their work trip and average commuting time of car commuters

Country	% of car commuters who encounter severe congestion	Average home-to-work time of car commuters (mins)
France	7	18
Germany	4	25
Netherlands	11	23
Italy	12	17
United Kingdom	19	17
European average	10	19

hours. Table 8.4 reports the percentage of daily car commuters in a number of countries that declare that they usually encounter many traffic jams on their way to work.

These figures again show the favourable congestion situation in France and highlight the relatively poor conditions in the UK. If we consider that congestion outside rush hours is relatively rare, the figures do not support the suggestion that congestion is very severe.

We may conclude that congestion is relatively rare when one considers the overall magnitude of the motorway network and the total amount of travel activity on the road system, even in the urbanized regions of Europe.

5. BEHAVIOURAL RESPONSES TO CONGESTION AND TO POLICIES

Responding Actors

The behaviour of three major classes of actors must be understood in order fully to understand the responses of actors to changing congestion and in order to design effective ways of coping with it. We distinguish between the behaviour of individual travellers who make their decisions concerning travel and its timing, modal choice, destination and so forth, and firms, which have a different set of choice variables. Each actor is concerned with different attributes of congestion and has a different set of optional responses.

The third major class of actor is governments, at all levels, which through their actions (or inaction) influence the attributes of congestion such as its size and temporal and spatial distribution. Presumably, governments devise policies as a result of studies and evaluations. Unfortunately, policy measures are often adopted as responses to short-term political pressures without the necessary groundwork. In this section we will discuss the response to changing congestion of only one set of the major players: the users whose mobility is impaired by congestion.

Travellers' Responses to Changing Congestion

The issue of response dynamics

When facing increasing congestion, individuals experience growing dissatisfaction, which may reach some threshold level that triggers a reassessment of the situation. Salomon and Mokhtarian (1997) have suggested that the following process takes place. A search is initiated when a certain level of dissatisfaction has been reached. Given the experience they have gained, namely prior adjustments to congestion, individuals identify the potential options for adjustment, evaluate them and choose a course of action, which is likely to reduce dissatisfaction, at least temporarily.

Once a choice has been made and some action is taken, dissatisfaction may be reduced at least for a while, but in the context of increasing congestion, a threshold point of dissatisfaction may be reached again, triggering another search for solutions. This time, previously adopted solutions may not be feasible or desirable. However, it is also possible to choose an alternative repeatedly, such as adjusting work-trip departure times, or changing routes. Consider the case in which some low-cost strategies were selected and subsequently a high-cost strategy, such as a residential relocation, was selected. In following rounds, the low-cost strategies may again be considered.

The innovation of this model lies in three elements. First, it addresses the issue of search initiation, through an identification of the *dynamics* of the process. Second, also through the dynamic perspective, it focuses on the individual's limited *choice set*; and third, it articulates the implications of the *lateral impacts* as factors which affect the behavioural response.

The dynamics of the process deserve significant attention to improve the likelihood of successful policy intervention. The timing of an adjustment decision, or deliberation about a decision, depends, among other things, on the history of such adjustments.

Understanding the issue of the time required for deliberation about change is important for policy-making considerations. This is likely to be a function of the transaction costs. Residential relocation is not a decision made on the spur of the moment, while route change may be. Thus, when a situation changes, or when a policy is introduced, there is a span of time in which each potential response may be employed. This is a very important point from a policy evaluation perspective. If a policy measure is evaluated before the range of likely responses has been adopted, premature conclusions may result.

The choice set

The 'universal set' includes a wide range of reasonable responses to changing congestion. Each individual may not face this full set, but a subset of these. The individual choice set is determined by constraints as not all responses will be available to a particular individual.

It is useful to classify the responses on the basis of the objective they fulfil for the individual. Based on Mokhtarian and Salomon (1997), the following range of responses should be considered

1. *Accept travel costs*: This 'do-nothing' situation seems to be a prevailing response. It may indicate that despite the public and political grievance about congestion, it may not be as severe a problem as commonly believed. In economic terms, it implies that the costs of adopting any other response strategy are greater than the costs of congestion to the individual.
2. *Reduce travel costs*: The automobile and car gadget manufacturers seem to cater to the frustrated driver by offering an increasingly pleasant and functional 'commuting environment': air-conditioning, a quality music system, a cellular telephone, and other elements of comfort make the time spent travelling by auto more acceptable.
3. *Adapt departure time*: This strategy can reduce travel time, if the peak period is relatively narrow. It will be less effective in those areas where congestion prevails for many hours continuously. Constraints such as rigid work schedules or needing to drive family members limit the ability to adopt this response.
4. *Change route*: By changing to a route with less stop-and-go traffic, the traveller may reduce commuting stress even though the new route may be longer or slower.
5. *Buy time*: By paying congestion tolls one can buy travel time, while paying parking fees may reduce access time. A popular strategy to compensate for time lost in travelling is buying the time of others, such as baby-sitters, household help or support services at the workplace. Investing in technologies for the home that increase productivity is another way of buying time. So extra travel time is compensated by time gains elsewhere (activities at home) at a certain cost.
6. *Temporal changes (flextime, compressed work week, and changes from full- to part-time jobs)*: Temporal changes allow diversion of trips from peak periods to other periods, either by the adoption of flextime or by adopting ten-hour four day work weeks.

7. *Change mode*: Switching to other, more efficient modes of travel is often the solution suggested by transportation professionals, environmentalists and politicians. However, based on experience, the success of that particular approach is limited to situations where congestion is very severe and shared-ride modes are competitive in time and cost to the automobile (for example in CBD-bound trips where parking is limited and costly).
8. *Telework (telecommute) from home or from a local work centre*: Alternative work arrangements which allow flexibility not only in time, but also in space, facilitate responses which allow the individual to avoid congestion.
9. *Relocation of workplace or home*: Avoiding congestion by locational adjustments is an option for long-term response. It can either reduce distance or facilitate travel on routes which do not suffer from congestion.
10. *Start a home-based business*: This strategy entails costs for the individual along with potential benefits like monetary gain, time, lower stress (greater control of one's work), and convenience (schedule flexibility).
11. *Quit work*: This response carries a monetary cost even greater than that of strategy 10. If the motivation to quit work is predominantly the stress of congestion, the result is likely to be deep frustration. Quitting work, which was mentioned earlier as a radical act, may in fact be quite common. We suggest that many people who do not work are those for whom the given (mostly time) costs of congestion have exceeded the costs of other responses and the benefits of work. This may be more common among women than men.

The list is ranked on the likely frequency of responses, but it also identifies three types of relevant strategies: responses which *maintain* the current level of travelling, by making travel cheaper or more convenient; responses which *reduce* travel; and *life-style/locational* changes. From a policy perspective, the latter two groups are of interest although the locational changes may result in undesired effects on congestion.

As travellers are assumed to be utility maximizers and not cost minimizers, they tend to explore the possible options for adjustment on the basis of what is good for them, which may not coincide with a societal perspective. Thus, when conditions change, individuals are likely to exhibit evasive behaviour, namely they will try to identify and adopt those options that are least onerous. By contrast, when constraints are relaxed, such as in the case of highway expansion, individuals may exhibit expansive behaviour, thus improving their relative position. The return to the peak phenomenon observed in some cases (see below) is an example of expansive behaviour. However, when road pricing is introduced, it may entail a variety of life style and locational changes, rather than the often-expected modal shift to public transport.

Few careful analyses of responses to changes in congestion have been performed. In most cases, research efforts focus on evaluating the effectiveness of a particular policy measure and do not monitor the wide range of options which individuals may consider and adopt. Some empirical evidence of adaptions to changing congestion can be found in Stern et al. (1995).

6. ADDRESSING CONGESTION: POLICY MAKING AND POLICY TAKING

With congestion being an important item on both the public and policy makers' agenda, it receives much attention in many countries. A search for policy measures that would curtail congestion is evident in the abundance of professional literature and popular debates in most of the developed world.

Beyond the wide consensus that 'something should be done', a wide range of views is brought forward and debated. This debate refers to the definition of *objectives*. What is the desired level of congestion? The more popular position is that congestion should be minimized, namely nullified, while from a societal perspective, the question is what level of congestion is appropriate. On the other hand, the question of the means may be considered: what types of policies can and should be pursued? Here, in a nutshell, the dispute is between promoters of effective policies which may be politically costly and policies which are politically attractive, but less effective or even ineffective.

This section opens with a discussion of the objectives, namely questioning the issue of a desired level of congestion. Does it exist? If so, how is it to be identified? Then, a brief review of policy approaches and an assessment of the potential policies in Europe is given.

A Desired Level of Congestion? An Economic Approach

Congestion imposes costs in a number of different forms. First and foremost, from a political perspective, are the personal costs incurred by the myriads of individuals whose travel times are prolonged due to congestion. Second are the costs incurred by society as a whole. This includes the uncompensated loss of time of individuals as well as the greater costs of friction in production systems. As congestion is also a very unstable situation, the travel-time reliability during congestion is low. This imposes additional costs on individual users, who have to allocate sufficient time to account for uncertainty in expected arrival times, as well as on firms, which need to adjust for lower reliability by adding, for example, to their inventories. Estimates in the USA have suggested a loss of $38 billion in 1988. In Europe, congestion costs are estimated at about 2 per cent of GDP (Kinnock, 1995).

If the costs are so high, something 'ought to be done'. But what exactly is the objective to be attained? The question of the *desired level of congestion* is only rarely addressed.

The suggestion that congestion should be minimized is dubious. It is not possible to design and implement a transportation system which will not experience congestion. That level of zero congestion can technically be accomplished only through an unreasonable investment in construction. For long periods each day this system will be underutilized, as will be the funds invested in its construction. From a societal perspective there is a desired level of social costs.

The desired level of congestion is that level which optimizes the social costs and benefits of the various elements involved while taking developments over time into consideration. The costs and benefits that should be considered include capital necessary for construction, maintenance costs, environmental costs (air pollution, noise and the value of land), safety costs and benefits, and travel-time losses and gains. It is clear that there is a trade-off between capital investment and congestion. Accepting higher levels of congestion will reduce construction and maintenance costs (fewer roads, fewer lanes), but the costs of travel-time losses will increase.

For roughly ten years, the approach adopted in the Netherlands has explicitly considered congestion in the design procedure for trunk roads. It was recognized that having congestion in the network is not all bad if it is contained to a desired level. The quality-of-flow criterion in use nowadays in the Netherlands for the design of motorways is the *probability of congestion*. This measure expresses, for a particular road section, the percentage of daily users of that section that experience a queue (as shown in Figure 8.4).

Compared to the classical criterion of speed, the probability of congestion adds two important criteria for quality of flow: travel-time loss, namely excess travel time; and reliability of travel time.

Economic analyses have shown that (in 1990) the optimal congestion level for the Dutch trunk road network is equal to a congestion probability of about 2 per cent (Stembord 1991). This optimal level means that, on average, over a long period, 2 per cent of daily traffic on a road encounters congestion to some degree. This needs some further clarification. If we assume that all congestion only takes place in the two peak hours, one in the morning, the other in the afternoon, each of which carries 10 per cent of daily traffic, then this 2 per cent means that there is a 10 per cent chance for peak-hour travellers to get into a queue on a working day (off-peak travellers will not meet congestion at all). That is only a few minutes' delay in a queue once a

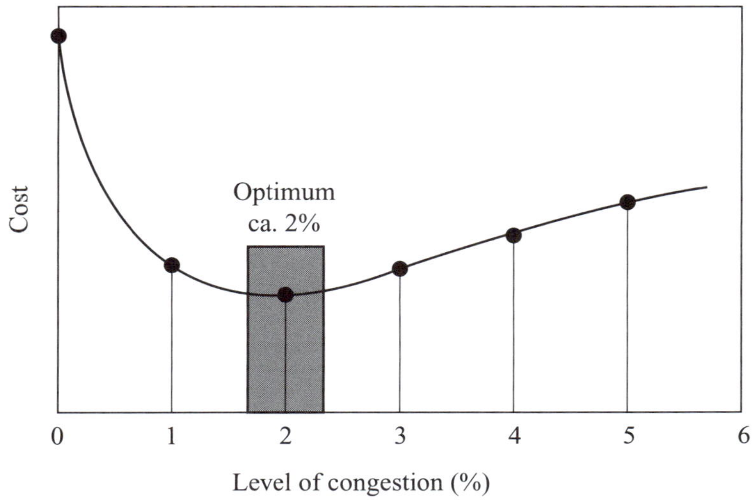

Figure 8.4 *Social costs of road infrastructure provision in relation to accepted congestion levels in the Netherlands*

fortnight. If we had, however, a road section with a congestion probability of, for example, 20 per cent (such cases do occur), this would describe a structural bottleneck with recurrent congestion during peak hours with delays of 20 minutes or more each working day.

The dimensioning of the roads and the capacity calculations are nowadays based on this 2 per cent congestion standard. It is considered an economic optimum.

Policy Approaches

It is possible to identify at least three historical periods in which policy measures to curb congestion have emerged from very different assumptions about the nature of the problem. Initially, and through the mid-1960s, the principal tool was expansion of infrastructure: more roads were built to accommodate demand. Later, there was a shift toward improved management of the available infrastructure. This was the Transportation Systems Management (TSM) period which prevailed during the 1970s, and TSM is still a relevant tool. However, TSM is also limited in its potential contribution, and in the early 1980s there was an increasing realization that altering human behaviour is the next necessary step. This led to the development and implementation of Transportation Demand Management (TDM) strategies, involving a wide range of policies to reduce dependence on the drive-alone automobile.

While the first two periods can be characterized as emphasizing supply-side measures, the third is by definition designed to affect demand. Supply-side measures which cater to accommodating demand are likely to be positively received by users (albeit not necessarily by non-users, who may be the very same individuals when they are not behind the steering wheel). Politically, measures which infringe on constituents' personal behaviour (and freedom) are considered undesirable, and therefore, according to Altshuler (1979), policy makers refrain from implementing policies which have direct negative impacts on users, such as those directed at modifying demand. Rather, where possible, policy makers will prefer a policy that 'looks good' even if its effectiveness may be limited.

The case of road pricing, widely advocated by transportation professionals as a promising congestion management policy, but so rarely applied, is a clear example of a policy that directly affects constituents' pocketbooks (Emmerink et al., 1994; Jones, 1991; Grieco and Jones, 1994; Wachs, 1994).

Supply-side and demand-side interventions differ in another respect, which is important in the current context. Generally, the direction of behavioural response to supply-side measures can be expected to conform to that anticipated by the policy makers, and the question is whether levels of adoption will be lower (as is often the case for ridership on a new transit service) or higher (as when the release of latent demand triggers nearly immediate congestion on a new facility) than forecast. However, in the case of demand-side measures, the individual is confronted with a situation that imposes a constraint. In this case, new 'outlets' are likely to be sought, and innovation may generate new, possibly unexpected, responses, as described below.

From a public policy perspective, congestion mitigation strategies can be classified into five groups, differing in the nature of intervention assumed: regulation; planning; economic; technological; and educational.

Regulation includes a wide variety of measures, many of which are at the disposal of local agencies, through which the policy maker exercises some power to alter behaviour of consumers, that is, transport system users. This may include parking restrictions, changes in schedules of work and schools, and the like. Regulations are not politically attractive and are likely to be economically inefficient, but their major advantage is that they are relatively easy to implement and consequently, if they are effective, will deliver the benefits in the short term.

Planning includes a variety of measures intended to change the physical environment by changing the spatial relationship of opportunities. Land-use changes which alter the density or the mix of land uses are often suggested to enable a greater reliance on non-motorized travel, and hence a reduction of travel by car. On the other hand, the planning (and implementation) of transport infrastructure facilities also changes the relative positioning of opportunities and consequently is likely to affect travel patterns. Using planning strategies to contain congestion is a long-term policy, and its prospects are widely disputed (Handy, 1997; Breheney, 1995).

Economic measures are widely considered to be effective policy instruments as they send the users unambiguous signals as to the desired change of behaviour. Some European countries have begun to implement such measures as congestion pricing or, more generally, road pricing. Pricing of central city parking to reflect the externalities is also a possible measure. It is expected that with the introduction of electronic toll collection systems, there will be growing interest in economic measures, despite the widespread opposition by the public and elected officials, who view it as another tax.

Technology-based approaches include a variety of measures which may improve the management of transport systems, including improvements in the management of road capacity. Intelligent Transport Systems (ITS) provide a range of options to act both upon the supply side and upon the demand side.

Education is the often-forgotten policy measure. Many policy makers and the public at large are often unaware of the nature of the congestion problem and, moreover, have unrealistic expectations with regard to its 'solution'. Hence education which is geared to explain the nature of the problem and the implications of various policies is important. However, the effectiveness of education, if any, is realized in very long time horizons, and that is probably the main reason for its lack of consideration as a congestion mitigation policy.

Figure 8.5 repeats the structure presented in Figure 8.1 above, but focuses on the policy side. It complements Figure 8.1 by showing how congestion mitigation policies can be enacted to affect various factors which cause congestion.

7. CONCLUSIONS

The analysis presented in this chapter gives rise to the following conclusions.

The Notion and Extent of Congestion

A clear, unambiguous and widely accepted definition of congestion and how it should be measured is not available. Systematic data collection on road congestion is absent (with the Netherlands as an exception). Realistic and comparable figures about the extent of congestion and its costs are therefore lacking (with the exception of the ECIS study). This hampers a valid international comparison of congestion conditions in European states and regions. Comparative studies can potentially provide insights into the different causes

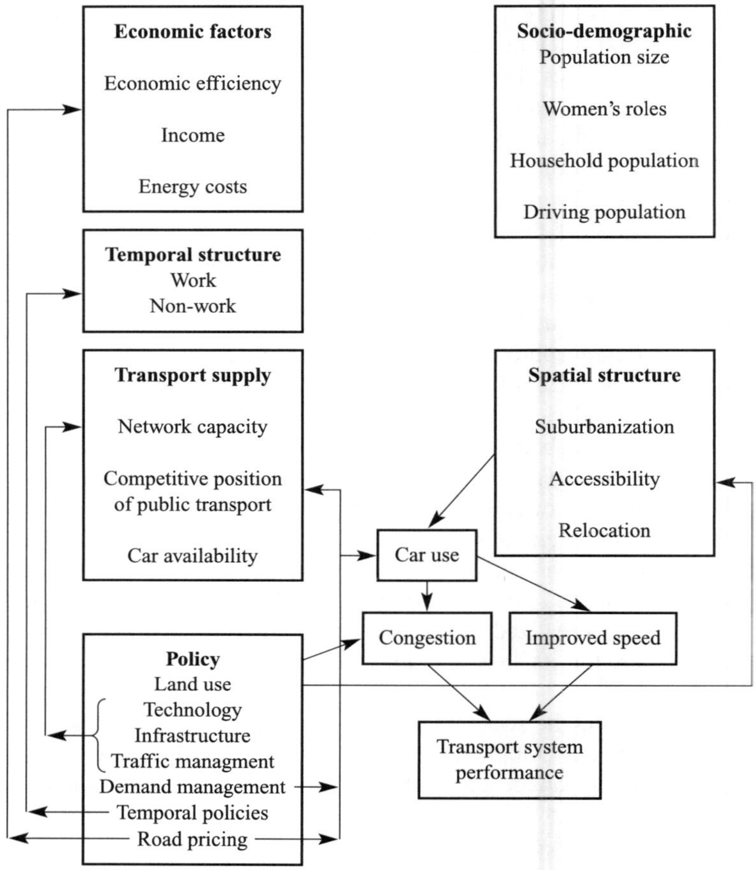

Figure 8.5 Influence of congestion mitigation policies on congestion factors

of congestion, under varying background conditions and, more importantly, lessons about the effectiveness of policy approaches. The figures presently available on the extent of congestion in Europe, as well as those used in many official European documents (EC, OECD, CEMT), lack a valid foundation and offer little sound basis for decision making.

The Use of Congestion Measures as Quality Indicators

Congestion figures, as such (for example queue length, congestion duration, excess time, excess costs), are poor indicators of network quality or trip quality. Congestion is not necessarily a sign of a poorly designed network or of an unacceptable quality of flow. From an economic perspective, there exists an optimal level of congestion in transportation systems which depends on local circumstances, such as construction costs, and travellers' value of time as well as the weight of transport and environmental considerations relative to other social problems.

The Spread of Congestion Levels

If measured at an aggregate (network) level (for example total excess travel hours, total queue lengths), congestion has increased constantly over the last decade, at a rate similar to that of motorway usage (on average 4 per cent per annum). Much of this aggregate growth is a spread of peak levels in time and space. This is a useful indicator from a public policy point of view. With respect to the travellers' choice behaviour, however, the congestion as experienced by the individual user, measured directly, remains quite constant.

The European Dimension of Congestion

The scale of the congestion problem is not European, nor is it national. It is also not a typical characteristic of major long-distance thoroughfares or border crossings. Road traffic congestion is an urban and regional (metropolitan) problem. It manifests itself predominantly in and around high-density conurbations. International comparisons, therefore, should rely on regional data rather than national-level data.

The True Costs of Congestion

Given the absence of unambiguous congestion figures, and the spreading of congestion in time and space, there is a widely accepted expectation of an imminent catastrophe or breakdown of the system. This view seems to be exaggerated and the costs of congestion, as shown in various official publications, may be overestimated. Congestion on main roads appears to be a local problem that affects a limited number of travellers. Recalculated to national level figures the excess travel time is most probably less than 2 per cent of total travel time and amounts to perhaps 0.3 per cent of GDP.

Variation in the Distribution of Congestion in Europe

Road traffic congestion is evident in many different regions of Europe. But congestion levels, growth rates and distributions differ widely between countries and regions. These differences can be attributed to variations in the underlying factors that generate congestion, particularly spatial patterns of land use and network conditions. The most important common factor is population density. In high-density areas, space for sufficient road infrastructures is scarce, whereas the density of potential users is high.

The Response to Congestion

When facing changing congestion levels, users of the networks (individuals and firms) have a wide range of behavioural responses at their disposal. Some responses are short-term travel adjustments while others are long-term, locational and life-style changes. A variety of behavioural changes is evident when policies to relieve congestion are implemented: shifts in time, route and mode diversion and a change of trip-making behaviour are among the common responses. They reflect mostly shifts in existing demand. New transport demand as a result of congestion relief appears to be very limited.

The Limits of Congestion

Total demand for road transport is not unlimited. Population growth and increased speeds due to increased welfare were the predominant factors for road transport consumption. The growth in total road traffic kilometrage is likely to diminish as populations grow at a slower pace and travel-time budgets of travellers are increasingly binding, giving rise to modification of activities in lieu of increased trip making.

Investment in Congestion Relief

Lack of investment in road infrastructure is a major factor in congestion growth in Europe. In most countries the lion's share of national infrastructure investment is in rail, predominantly for long-distance connections. These will hardly contribute to road congestion relief because they serve thin, long-distance travel flows. A serious problem is that investments in long-distance rail are at the expense of short-distance public transport networks and road investments as well. The contribution of urban and regional public transport investments to congestion relief is limited because of inherent system characteristics of public transport (such as spatial coverage and service qualities).

8. RECOMMENDATIONS

Based on the analysis and conclusions presented above, a number of pertinent recommendations are warranted.

Statistics on Congestion

In view of the dominant role of congestion in public policy making, there is a need for development of clear, unambiguous operational definitions for congestion measures and measurement methods suited for cross-regional comparison. A system of European-wide statistics on congestion should be established. A distinction should be made between aggregate (network-wide) measures and individual (traveller-related) indicators. Such statistics will improve information provision about congestion conditions and their societal implications, and will thus enhance transport policy decision making, in particular with regard to investments in infrastructure, pricing and public transport.

Optimum Level of Congestion

The notion that there is an optimum non-zero level of congestion should be developed further and should be communicated to opinion leaders, politicians and interest groups. An approach should be developed to determine the optimum congestion level that can be used in road network planning and design.

Need for Balanced Spatial Development

Long-term solutions to congestion must be directed at curbing the growing travel demand and travel distances by a balanced development of infrastructural networks and spatial distribution of activities. New concepts of spatial configurations of settlements need to be developed and assessed. It is still unclear how spatial structure affects changes in behaviour, but there is no doubt that density is closely related to the efficiency of different modes of travel. Thus careful examination of these relationships is warranted.

Public Transport is an Ineffective Congestion Relief Measure

Development and stimulation of alternative modes (such as public transport) as a means to tackle congestion is not a very effective policy approach in general, except for specific high-density corridors. This is due to their inherent system characteristics. Rail transport can only serve very limited segments of the travel market. Public transport must play an important function in offering services in dense areas and corridors and giving mobility opportunities to the carless.

Only the Road Can Relieve the Road (Gerondeau, 1997)

Addressing congestion from the supply side is most effective because it intervenes in the road system itself instead of suggesting alternative modes. Such an approach includes offering extra capacity by widening roads, building buffers to reduce secondary congestion, increasing road capacity and capacity utilization by dynamic traffic management and including demand management techniques, such as congestion pricing. This does not imply that the road system should be indiscriminately expanded. Other considerations, such as environmental quality, and social and spatial impacts, must also be taken into account. However, the hope that public transport can solve congestion is probably an illusion.

Need for High-quality Roads

Given the forthcoming demographic and economic developments, there seems no alternative but to upgrade the road infrastructure significantly, both in quantitative (additional capacity) and qualitative (environmental) respects. New roads as well as upgrading of existing roads must meet high environmental and aesthetic standards. Because congestion-prone areas are characterized by a scarcity of space, and a vulnerable natural and manmade environment (noise, aesthetics, and so on), costly solutions are inevitable. These may include, in extreme cases, underground or deepened roads, double stack roads, roads with covering and special (double-layered) tunnels for private cars and trucks. In addition, investments are needed in high-quality transfer facilities at the fringes of cities for travellers going to city centres by public transport or other high-occupancy vehicles. In view of the dynamics of the process of congestion building, such high-quality roads may not need to increase speeds, but ensure flow at reasonable levels of service.

REFERENCES

Altshuler, A. (1979), *The Urban Transportation System*, Cambridge, MA: MIT Press.

Bovy, P.H.L. and I. Salomon (1999), 'The Spread of Congestion in Europe', *ECMT Round Table*, **110**, 85–153.

Bukold, S. (1997), *Bottlenecks in European Infrastructure*, Rotterdam: ECIS.

DHV/Colquhoun (1991), *The Cost of Inadequate Transport Infrastructure in Europe*, Report to European Parliament.

ECMT (1993), *Survey Procedure and Results by Country*, Paris, November.

ECMT (1995), *European Transport Trends and Infrastructural Needs*, Paris.

Emmerink, R.H.M, P. Nijkamp and P. Rietveld (1994), 'How Feasible is Congestion Pricing?', Tinbergen Institute, pp. 94–62.

Gerondeau, C. (1997), *Transport in Europe*, London: Artech House.
Gordon, P. and H. Richardson (1991), 'The Commuting Paradox', *Journal of the American Planning Association*, **57**(4), 416–20.
Grieco, M. and P.M. Jones (1994), 'A Change in the Policy Climate? Current European Perspectives on Road Pricing', *Urban Studies*, **31**(9), 1517–32.
Handy, S. (1997), 'Travel Behaviour–Land Use Interactions: An Overview and Assesssment of the Research', Presented at the IATBR Meeting, Austin, TX.
Hendriks, F. et al. (1997), *Infrastructureel investeringsbeleid in vergelijkend perspectief* (in Dutch), (Infrastructure investment policy in a comparative perspective: a transport policy analysis of the Randstad, Ruhr area and Flanders city ring). Tilburg: Catholic University/TNO-Inro.
Hilbers, H.D. and E.J. Verroen (1996), *An International Comparison of Accessibility and Congestion Problems of Urban Areas: Can We Still Compete with Our Neighbors?*, PTRC.
Hilbers, H.J. et al. (1997), *Infrastructuur en mobiliteit in de Randstad* (in Dutch), (Infrastructure and mobility in The Randstad, the Ruhr Area and the Antwerp–Brussels–Ghent region). Delft: TNO-Inro.
IMD/World Economic Forum (1996), *Global Competitiveness Report*, Genova: IMD.
Jones, P. (1991), 'Gaining Public Support for Road Pricing through a Package Approach, Traffic Engineering and Control', April.
Kinnock, N. (1995), *Towards Fair and Efficient Pricing in Transport*, Brussels: EC.
Ministry of Transport (1996), *An International Comparative Study on Infrastructure*, The Hague: SDU Publishers.
NEA (1996), *ECIS Study: Bottlenecks in European Transport Networks; Road Transport*, Rijswijk: ECIS.
Pucher, J. and C. Lefevre (1996), *The Urban Transportation Crisis in Europe and North America*, London: Macmillan.
Quinet, E. (1994), 'The Social Costs of Transport', in ECMT, *Internalising the Social Costs of Transport*, Paris: OECD, pp. 31–75.
Salomon, I., P.H.L. Bovy and J.-P. Orfeuil (eds) (1993), *A Billion Trips a Day: Tradition and Transition in European Travel Patterns*, Dordrecht: Kluwer Academic Publishers.
Salomon, I. and P. Mokhtarian (1997), 'Coping with Congestion: Reconciling Behavioural Responses and Policy Analysis', *Transportation Research D*, **2**(2), 107–23.
Stembord, H.L. (1991), 'Quality of Service on the Main Road Network in the Netherlands', in U. Brannolte (ed.), *Highway Capacity and Level of Service*, Rotterdam: Balkema, pp. 357–65.
Stern, E., P. Bovy and M. Tacken (1995), 'Traffic Congestion and Behavioural Reaction', European research conference on European transport and communication networks: Policies on European networks, Espinho, Portugal, 17–23 April.
Transroute ISIS, Heusch-Boesefeldt and A.T. Kearney (1992), *EC Motorway Network Perspectives*, Study for CEC-DG-VII.
Wachs, M. (1994), 'Will Congestion Pricing ever be Adopted?' *Access*, **4**, 15–19.
Westland, D. (1997), *The Gattis Hypothesis Tested on Dutch Motorway Bottlenecks*, Delft University of Technology, Faculty of Civil Engineering.
WP5 (June 1994), *Report on the Methodological Basis for the Definition of Common Criteria Regarding Bottlenecks, Missing Links, and Quality of Service of Transport Infrastructure Networks*, Bonn.

9. Geographical classification of trips as a tool for policy assessment

Jean-Pierre Orfeuil

1. INTRODUCTION

Traffic, and especially car traffic, is increasing in most of the European countries at an annual growth rate of between 1 and 4 per cent, depending on the country. The majority of car traffic is generated by everyday life activities and occurs where people live, that is in urban areas and their environs.[1] Many authors have pointed to a contradiction between these annual growth rates and the fact that road networks cannot expand at the same speed, especially in urban areas where land availability is very limited. These authors have predicted major congestion as a result of this contradiction.

Congestion does exist, but, at the same time, one can observe that in France the mean car speed continues to increase. It is important to clarify and understand this observation.

A second point to investigate is the dependence of traffic growth on the suburbanization process: should we consider this growth as the result of urban sprawl only, or is it the result of other factors, such as a decreasing attention paid to distances in the selection of our activity locations? These questions are addressed in section 2, with census data describing the relative positioning of home and job locations.

A third point is traffic-calming strategies and their ability to reduce car traffic. Can we use a geographical analysis of travel patterns to estimate the potential benefits of these strategies at a national level? Section 3 deals with this question.

Urban sprawl is often associated with car dependence, and, as a result, decreasing air quality and increasing unsustainable energy consumption trends. Can we evaluate the energy consumption and the exhaust associated with the daily mobility of citizens, and compare these levels in different zones? Section 4 considers this question.

Urban sprawl is also associated with rational decision making, such as the search for a better environment and for affordable residences. On the other hand, these benefits may entail higher transport costs. Section 5 deals with the simultaneous evaluation of housing and transport costs according to the area of residence.

Addressing these problems implies the use of some type of geographical analysis of mobility patterns. This approach was used to compare mobility patterns in the Netherlands and in France and was presented by the author in *A Billion Trips a Day* (Salomon et al., 1993). This chapter presents examples of studies conducted with geographical concepts as a basis for analysis.

2. METHODOLOGY

This chapter aims to show that using geographical analysis of trips and mobility may improve our understanding of the problems raised by exploding mobility and of the possible successes and failures of development policies. The chapter relies largely on empirical evidence from the data sources described in the next section.

Data

Sections 1 and 3 are based on the National Personal Travel Survey (NPTS) conducted in 1994. The size of the survey sample is 15 000 households. The survey covers several topics. In the framework of this chapter, travel diaries completed by respondents over five years old are used.

Section 2 is based on the census conducted in 1975 and 1990, and especially on the home and work locations of the employed population. Sections 4 and 5 are based on a travel survey conducted in the Ile-de-France region in 1991. The sample size is again around 15 000. Individual data are used and then aggregated by zone or type of geographical link.

Geographical Concepts

France is a low-density country compared to those of the European core. Its area is traditionally characterized by the French Statistical Institute according to two different conceptual frameworks:

1. *Urban units* as opposed to rural areas. Urban units are places incorporating at least 2000 inhabitants, under a criterion of a contiguous built-up area. This is a purely morphological concept, which had a clear significance when walking was the dominant transportation mode.
2. *Urban zones* as opposed to remote rural zones (which account for 25 per cent of the French area but only 4 per cent of its population). Small urban zones are the zone of influence of small urban kernels, with a

minimum of administrative and commercial facilities. For the larger ones, the concept corresponds to metropolitan areas. For the purpose of the differential analysis, we usually distinguish five types of zones:

- the Ile-de-France metropolitan area;
- 32 metropolitan areas with at least 300 000 inhabitants;
- 155 metropolitan areas with a population between 50 000 and 300 000 inhabitants;
- the 415 other zones; and
- the remote rural areas.

Within each zone, the largest urban unit is identified and the main city considered the centre of the conurbation (Paris, Toulouse, Bordeaux), with the rest of that largest urban unit considered as the *urban suburbs* of that unit. In some analyses, inner suburbs and outer suburbs are distinguished.

The parts of the metropolitan areas which do not belong to the largest urban unit are considered periphery. Within that space surrounding the main city, it may be useful in some analyses to distinguish 'urban periphery', which in some cases can be considered as 'secondary poles', and rural periphery. Trips can now be classified according to their origins and destinations, with a first split between trips that link two different zones (external trips) and trips that link two areas within a conurbation, for example periphery to suburbs or central city to suburbs, and so on.

Another distinction is made for the estimation of traffic-calming strategies. In these analyses, trips of less than three kilometres are considered separately from longer trips in order to take into account the potential of cycling and walking.

A Broad View of Trip Patterns and their Evolution over the Study Period

The geographical patterns of local trips are given in Table 9.1. Two observations are prominent:

- Trips and trip mileage within a central city do not increase over the period. The greatest mileage growth is for external trips and trips between peripheries and suburbs.
- Trips with an origin and/or a destination in the central city represent 43 per cent of the motorized mobility, and only one-third of the total mileage, as they are the shorter trips. Another third is made up of

Table 9.1 Geographical patterns of traffic flows related to local mobility in France (1994)

	Trips/day			Trip-km/day			
	1982 (mn)	1994 (mn)	% (1994)	1982 (mn)	1994 (mn)	% (1994)	Growth 1982–94 (%)
Within a central city	27.8	27.6	21.4	94.5	89.1	7.5	–6
To/from central city	19.0	27.8	21.6	215.8	318.4	26.8	+48
Internal to suburbs	23.0	27.7	21.5	125.4	168.7	14.2	+35
Internal to peripheries	25.7	26.4	20.5	108.3	134.6	11.3	+24
Exchanges suburbs/ peripheries	2.3	5.2	4.0	47.2	85.6	7.3	+81
Exchanges between conurbations	10.7	14.3	11.1	223.1	390.0	32.8	+75
Total	108.5	129	100	814.4	1186.4	100	+46

Note: Field: all 'local' trips (less than 100 km), all modes except walking, working day.

Source: Gallez and Orfeuil (1997a).

tangential trips – within suburbs or peripheries – and the last third comprises exchanges between conurbations. This finding suggests that classical traffic-calming strategies, which are mainly focused on city centres, may be efficient on only a part of the local mobility market. Furthermore, this part is not the one which is experiencing the highest growth.

3. COMMUTING PATTERN TRENDS

Data used in this section come from the 1975 and 1990 census (Massot and Orfeuil, 1995). They concern the working population only.

Analysis

The development of home and job locations of the working population
As a mean, central cities, which constitute 5.2 per cent of the area of the metropolitan areas, continue to concentrate a large part (44.1 per cent) of the jobs. However, these areas have lost 5 per cent of the total job supply and 17 per cent of the working population between the two censuses. That negative dynamic affects inner suburbs as well. The main gainers in job locations are the outer suburbs (25 per cent increase in jobs) and, in a lower

proportion, the nearest periphery. The main increase in housing is in the nearest periphery. The distant periphery has lost jobs and gained working population.

The degree of these changes depends on the size of the conurbation. In the smallest ones, urban sprawl affects housing locations only, with jobs remaining concentrated in the urban kernel. The underlying reason for this change is the decline of agriculture, local shops and traditional village activities. The decentralization of workplaces appears in conurbations of more than 300 000 inhabitants and grows with the size of the area, with the Ile-de-France region at the top. In all the cases, however, jobs retain a much higher degree of concentration than residences.

Outer suburbs, the zones with the highest growth in employment, are served by the orbital roads. These roads seem to be a major relocation factor for companies.

The changes in the distribution of work and home locations does not entirely explain the growth of commuting distances

During the study period, the mean distance of job locations to the city centre has grown by 5 per cent and the distance of home places to the city centre by 12 per cent. The greatest shifts are observed in the Ile-de-France area, with growth rates of 16 per cent and 19 per cent respectively for job and home locations. If these changes were the main explanation for the increase in commuting distances, this increase would be in the same order of magnitude. This is not the case, as the mean growth over the period is 56 per cent. The growth of commuting distances is less related to different dynamics in the distribution of work and home places than to the individual matching of homes and jobs. The underlying factors may be increased job mobility, job specialization, increased home stability, especially with regard to the growing number of home owners in France, and, to a lesser extent, the growth of multiactive households.

The patterns of accessing origins and destinations avoids the most congested areas

First, exchanges between urban areas have the highest growth rate. They use the intercity network, with travel speeds considerably higher than those that can be reached for trips within a conurbation.

Second, commutes to the city centres and inner suburbs are growing much less than commutes to outer suburbs. As a result, entries in the low-speed zones have the lowest level of growth. *Reverse commuting* (commuting from an inner zone to an outer zone) currently accounts for 10 per cent of trips within a conurbation and is not presently affected by congestion.

A small part of the population is responsible for a large amount of traffic

Despite the increase of the mean travel distance (from 7.4 to 11.5 km, a 56 per cent growth), 75 per cent of commuting trips are shorter than 10 km and only 4 per cent are greater than 40 km. Considering the total traffic, the 75 per cent of commuters with a travel distance less than 10 km account for 24 per cent of commuting traffic, expressed in passenger-kilometres, while the 4 per cent of those who commute the longer distances account for 45 per cent of the total traffic. This observation may explain the apparent contradiction between the diagnosis of travel experts who, observing the total traffic, see problems and the opinion of the public, which is generally that they do not experience large problems with traffic. Looking at traffic as a whole (which is, for example, the general basis for economic assessment of roadways) emphasizes the total distances travelled. Looking at people (which is the general basis for policy makers) emphasizes trips, and the conclusions that can be drawn are quite different. Even on the basis of objective observations, a difference between expert and public opinion may be understandable.

Policy Implications

- Orbital roads have been built around the main cities to avoid transit through urbanized areas. This policy aim is achieved, but these roads have been a major contributor to the decentralization of employment locations, thus making the trip market more car-oriented.
- Economists and traffic engineers refer to the total amount of traffic when they speak of flows, traffic and congestion, and ignore that a large part of the car traffic is generated by a small part of the population. Policy makers have to take into account that the majority of the population does not experience great problems, because its level of car use is low. This may be, among other reasons, an explanation for the problems encountered by the development of road pricing programmes.

4. THE USE OF TRIP GEOGRAPHY TO ESTIMATE THE POTENTIAL OF TRAFFIC CALMING MEASURES

Analysis

Environmental and quite often scientific literature on sustainable mobility focuses on mode and modal usage. Walking, cycling and public transport are considered the 'good' modes, that is, within the environment-friendly family, while the car is cast in the role of the 'bad' mode. The distribution of trips according to mode is of course one dimension of the mobility market, but not the only one. Walking and cycling are better suited to short trips and public transport is better suited to centripetal trips and trips within dense zones. The efficiency of the modes varies according to the origin and destination of the trips. Likewise, the potential for splitting between cars and a mode belonging to the ecological family depends on the types of origins and destinations. We have been asked recently to estimate the potential of car traffic reduction by classical traffic-calming measures: specifically, what reduction in car use (expressed in car mileage) may be expected from the development of new cycling, public transport, and parking regulation programmes as well as the effect of changes in home locations designed to curb car traffic?

Table 9.2 Maximum avoidable car traffic through traffic-calming measures (billion vehicle-km/year), current total in conurbations over 50 000 inhabitants: 115 BVK.

Measure	Billion vehicle-km/year
Trips within conurbation	
Walk and cycle promotion on short trips	4
Promotion of public transportation for radial routes	3
Promotion of public transportation on suburbs-to-suburbs market	0.5
Street parking management	7
Employers' parking management (reduction of supply: 1%/year over 20 years)	2
Trips external to conurbation	
Transfer to PT for the centre-to-centre market	2
Avoiding new urban sprawl	12

Source: Gallez and Orfeuil (1997a).

An estimation at the national level cannot be realized through a detailed model, taking into account the characteristics of the transport supply. On the other hand, a detailed presentation of the geographical structure of trip making (see Table 9.2) may help to approximate the structure of the market. Here the 1994 National Personal Travel Survey (NPTS) is used to characterize the geography of trip making and the current associated levels of car use for each of the four sizes of conurbations, considering separately intercity trips and trips within the conurbation.

Trips are classified according to length (less/more than 3 km) and origin/destination (from centre/suburbs/periphery to centre/suburbs/periphery for internal trips, between centres, including one centre at the origin or the destination, no centre as origin and destination for external trips).

For each geographical type of trip, the following information is computed:

- The total distance travelled by type of origin/destination and size of conurbation. This information provides the size of the different markets.
- The mode split of that distance: car as a driver; car as a passenger; public transport (urban/regional), walking and cycling. For public transport, the total level of demand is split between people without driving licences and people owning driving licences.
- The ratio of car speed to the speed of the potential competing modes.
- Other specific indicators are used, especially for short-distance trips, such as the share of trips which are included in multistage journeys or the share of trips with transport of loads.

It is clear that, at a national level, these indicators cannot be used to estimate the actual behavioural changes which would result from policy implementation. They may only be used to estimate the maximum potential of each measure under realistic policy options. For example, transfers from car to walking or cycling are considered as impossible over a certain distance threshold, parking policies are considered efficient only in city centres and some other densely populated areas; the potential for public transport development is considered different in small conurbations and large ones as well as on the links with city centres and on the other types of origin/destination. The ratio of speeds is considered as very limiting for transfers if its current value is over 2.5 for car speed relative to the competing mode. The estimation of the potential transfer must be considered as the upper limit for likely transfers. However, it appears to be limited (see Table 9.2) to less than 15 per cent of the current level of demand.

A change of the distribution of residences has been tested in parallel, under two different hypotheses. The first assumption is that the dynamics of suburbanization continues at the same pace as in the past. The alternative hypothesis assumes that new populations in large metropolitan areas are proportional to the current population of the different areas of the zones (that is, metropolization continues, but there is no progress or regression of suburbanization) over a 20-year period. In both cases, it is assumed that new residents adopt the current travel patterns of their neighbours. These simulations suggest the potential of reducing car traffic by 12 per cent after 20 years.

These results may be considered questionable, either from a methodological point of view or from the point of view of the basic assumption that there are no changes in the origins and destinations of trips. On the other hand, considering the current modal shares (84 per cent for car, 10 per cent for public transportation, 6 per cent for walking and cycling, in passenger-kilometres) cannot lead to very different estimates, if the origins/destinations are presumed to be unchanged.

Policy Implications

Classical traffic-calming measures probably have a huge potential for improving the quality of the environment in the denser parts of the city; however, these areas account for only one-third of local car traffic. If it were demonstrated that dramatic changes are necessary regarding car use (for example to mitigate greenhouse gas emissions or avoid new road constructions), then the only way to obtain the changes would be to work, shop, make private business visits or pursue leisure in a less extensive area. This observation helps to clarify certain hesitations regarding implementing traffic-calming measures. Specifically, traffic calming is efficient if it changes the structure of origins/destinations, which means very often the size of the markets, and as a result is likely to create winners and losers.

5. URBAN SPRAWL AND SUSTAINABILITY: ENERGY AND ENVIRONMENT BUDGETS AS AN INDICATOR

Analysis

A geographical analysis of local or national travel surveys, when combined with models which associate a given trip with a given level of energy consumption and a given level of exhaust emissions, can answer the following three types of question:

1. What is the level of petroleum consumption and CO_2 emissions of a typical person according to his/her area of residence, characterized for example according to the distance to the city centre or the density of the municipality of residence? This indicator is suited to global sustainability approaches.
2. What is the density of emissions (for instance g/km^2 of nitrogen oxides or volatile organic compounds) in a given zone? Even though air quality is largely formed at the regional level, it may be useful to know where the pollutants are emitted, because of the sensitivity of population to local emissions.
3. Who is responsible for emissions in a given location or type of location, for example a city centre?

To answer these questions, local travel surveys or national travel surveys can be used and complemented by information on light and heavy truck traffic. A trip is characterized by its origin and destination, distance, travel time and mode. When the mode is a car, different levels of consumption and emissions are used according to the type of car (gasoline/diesel) and its age (to take into account the different levels of regulation with which it complies). Finally, emissions of a given pollutant are computed as a function of distance, speed, and type of car emission technology. When public transport is used, its emissions are computed on the basis of its technology (electric/thermic) and its occupancy rate.

After these computations, two types of allocation processes may be undertaken:

1. The simplest one allocates the trip to the trip maker. The sum of all the consumption and emissions of the trip maker over the day constitutes his 'energy and emission budget'.
2. The second consists of allocating the trip or parts of the trip to zones, in order to evaluate the car density and the related exhausts to zones. For example, a trip from a periphery to a city centre has to be divided into traffic within periphery, traffic within suburbs and traffic within city centre. With a local travel survey, that operation may be undertaken through an assignment model. With a national travel survey, assignments may be estimated on the basis of relative sizes (area, radius, and so on) of the different zones. Some of the results of these approaches are presented below.

Urban sprawl is clearly a problem for CO_2 emissions

Energy and CO_2 budgets per head grow sharply (see Table 9.3) from city centre locations to peripheral areas. The growth may be observed either according to the distance to the city centre or to the reduction in density.[2]

The growth of CO_2 emissions for a given type of person (active man, active woman, housekeeper, retired, student and so on) has a similar general shape, and the mean ratio is from 1 to 3 from the central locations to the suburban ones. For commuters, the growth is more sensitive to distance to the city centre than to local density. For the others, the growth is more sensitive to the local densities as their mobility is more related to the local characteristics. As a result, new towns and secondary towns have lower levels of CO_2 emissions than the rural areas included in conurbations, even when they are located at the same distance from the city centre.

Dense parts of the city have the highest density of emissions

Allocating emissions to the zones where they take place is more useful for local pollutants. Table 9.4 shows that despite the growth of trips in peripher-

Table 9.3 CO_2 emissions for transport and travel behaviour in Ile-de-France in relation to density of municipality of residence

Inhabitants/km^2	CO_2	Travel time	Distance	% as a car driver
< 2 500	3 000	78	30	55
2 500–5 000	2 200	82	24	44
5 000–10 000	1 900	83	19	43
10 000–15 000	1 400	78	16	39
15 000–20 000	1 500	86	15	42
20 000–30 000	1 200	87	15	32
30 000 +	1 000	93	17	24

Notes: Density: number of inhabitants per km^2; CO_2: g/day/person; travel time per person per day in minutes; Distance in km per person per day; share of distance as a car driver in % of the total distance.

Source: Hivart and Gallez (1997).

Table 9.4 Road traffic and exhaust emissions, France (km^2/day)

	Veh.-km/ km^2	NO_x kg/km^2	Particulates kg/km^2
Large French cities outside the Ile-de-France	2 600	6.5	0.5
Centre	33 700	78.6	6.9
Suburbs	11 400	25.3	2.1
Periphery	1 400	3.8	0.3
Ile-de-France region	8 700	20.3	1.6
Paris city	133 400	297.3	24.3
Suburbs	42 100	88.0	6.7
Periphery	2 600	7.9	0.6

Source: Gallez and Orfeuil (1997b).

ies, central zones remain the place of very high levels of local densities of emissions, compared to suburbs or peripheries.

Policy Implications

Urban sprawl increases car use and CO_2 emissions for transport dramatically. If the CO_2 question becomes a political problem, policies directed toward more compact cities will be needed. They will have to be complemented by traffic-calming strategies in the denser parts to avoid the growth of other problems such as noise, pollution and overcrowding of local streets by cars.

6. HOUSING EXPENSES AND TRAVEL EXPENSES IN THE ILE-DE-FRANCE REGION

Analysis

Suburbanization and urban sprawl are frequently considered as resulting from the desire for larger residences with a better outdoor environment, with only implicit references to the structure of the housing market and its relative prices. Differences regarding mobility patterns are often analysed from the point of view of transport (mode share, distance travelled per day) and less often from the point of view of monetary expenses associated with the different locations.

We have merged transportation observations from the 1991 Ile-de-France transport survey and housing cost data from surveys on the housing market for two categories of people:

(a) tenants who rent a home (either a flat or a private house) in the private sector (as opposed to the public sector); and
(b) people who are in the process of paying off their mortgage to become home owners.

For these two categories, monthly housing costs have been used to divide the Ile-de-France region into nine categories of housing costs, from roughly 100 FF/m^2/month in the most expensive (and central) locations to 50 FF/m^2/month in the cheapest (and most remote) locations. The transportation observations have been converted in monetary costs on the basis of fares for public transport and the full cost (including depreciation, maintenance and repairs, insurance, fuel, and so on) for car usage. As a result, the housing and travel budgets may be computed, and their shares in the income of the household may be estimated.

Comparing the main features of the households and their expenses leads to the following conclusions:

- The number of persons per household grows regularly as one moves from the expensive zones to the cheap zones.
- The income per consumption unit decreases sharply from the expensive to the cheap zones. The consumption unit is defined according to

the 'Oxford scale' in which the first adult is equivalent to 1 unit, the other adults to 0.7 unit, the children to 0.5 unit.

- The available square metreage per person is fairly steady in the different zones. Moving further from the centre is a means for less affluent households to obtain as much living space as the other households, but is not associated with higher absolute levels of comfort.

As a result of home location and type of occupation (tenant or owner), four types of expenses may be analysed: the mean travel time per capita; the mean travel time for the head of the household; the housing cost of the household; and the travel costs generated by all the members of the household for everyday life (excluding long-distance trips, such as holidays). These expenses are summarized in Table 9.5 and discussed in greater detail below.

Table 9.5 Travel characteristics and shares in the family budgets of housing and travel expenses in relation to area of residence

	1	2	3	4	5	6	7	8	9	IdF
Travel time/person/day (mins)	81	86	87	87	84	82	82	78	81	82
Distance/person/day (km)	10	10	12	13	14	16	19	18	24	16
Share of housing costs in the budget (%)	30	27	26	24	29	27	28	29	29	27
Share of transport costs in the budget (%)	5	6	7	9	11	14	19	19	26	14

Note: Zones 1 to 9 are defined by reference to housing costs, from the most expensive (1= the western part of Paris) to the least expensive ones (the southeast border of the Île-de-France region).

Source: Polacchini and Orfeuil (1998).

Small variations are observed in the travel time budget per head of the household. Travel time budgets are a bit lower for tenants in the cheapest locations while the opposite trend may be observed, with significantly higher differences, for the heads of the households who are home owners. It appears that people who want to become home owners do accept higher travel-time budgets, especially when their budget only allows for a remote location.

Housing expenses represent a fairly steady portion of the total budget (around 26 per cent for tenants and 28 per cent for future home owners), with no significant levels of variation according to zones. This is probably related to the fact that landlords and banks apply strict rules regarding the share of housing expenses in the family budget (generally allowing no more than 30 per cent).

The range of variation in the share of monetary travel expenses in the household budget is very large in relation to the dominant use of public transport in expensive areas and to the growth of travel distances from expensive to cheap zones (as a mean from 10 to 24 km per capita, and from 13 to 34 km for the head of the household). The mean share is around 15 per cent for tenants and 19 per cent for future home owners; the lowest shares are observed in the expensive zones of the housing market (7 per cent for the two categories) while the highest shares are observed in the cheapest zones of the housing market (21 and 27 per cent respectively).

Policy Implications

The structure of the housing market, the public incentives to home ownership and the precautions taken by landlords and public banks lead low- and medium-income families to residential zones where travel expenses may be quite high. A more balanced approach should take into account both the housing costs and an estimation of travel costs necessary for everyday life. This calculation would lead to different decisions of location, and different levels of car ownership and car traffic. In France public incentives are directed more towards home ownership than tenancy. This policy leads again to higher travel expenses and car traffic.

Alleviating road congestion through transportation price increases would pose severe problems to the households who have been driven to remote and cheap locations. More than half of their household budgets is devoted to housing and transport, while central city dwellers devote only one-third of their budgets to these two functions.

It should be added that we have conducted a specific piece of research devoted to double-income households, with at least two jobs in the household. It appears clearly that, in the cheapest zones, the two workers depart in the same direction each morning. This finding suggests that the 'choice' of home location is mainly based on considerations of affordability.

6. CONCLUSION

The analysis of the geographical patterns of mobility seems to be a prerequisite to identifying different mobility markets and to designing sustainable transport policies without unsustainable hopes. There is a potential for modal shifts in the denser parts of conurbations, but this market accounts for only one-third of total mileage. In the other markets, where classical traffic-calming strategies would be less efficient, pricing may be efficient, but low-income people driven to peripheries by housing markets could suffer. A key variable of sustainable transport appears to be land-use control and housing policies taking into account the undesirable effects on car traffic.

ACKNOWLEDGEMENTS

The author thanks the French institutions that have supported this research (ADEME, CNRS, DTT, DREIF), the colleagues with whom they have been conducted (Caroline Gallez, Marie-Hélène Massot and Annarita Polacchini), and the anonymous referee who has provided valuable comments to improve the chapter. The author would also like to note that this chapter was researched and written largely in his position as research director at the National Institute of Research for Transport and Safety (INRETS).

NOTES

1. In France the share of car traffic generated by long-distance trips (over 100 km) is less than one-third of the total car traffic.
2. Detailed results are available for Bordeaux, Grenoble, Ile-de-France, Lille urban regions and at the national level.

REFERENCES

Breheny, M. (1995), 'The Compact City and Transport Energy Consumption', *Trans. Inst. Br. Geogr. NS*, **20**, 81–101.

Gallez, C. and Orfeuil, J.P. (1997a), *Politiques locales et maîtrise des déplacements en automobile: une analyse des potentiels de régulation*, Paris: INRETS.

Gallez, C. and Orfeuil, J.P. (1997b), *Stratégies de réduction de la pollution de l'air lié aux transports*, Paris: INRETS.

Hivert, L. and Gallez, C. (1997), *Les Budgets Énergie Environnement des déplacements*, Paris: INRETS.

Massot, M.H. and Orfeuil, J.P. (1995), *Espaces de vie, espaces de travail: 15 ans d'évolution*, Paris: INRETS.

Polacchini, A. and Orfeuil, J.P. (1998), *Budget logement et budget transport des ménages en Ile-de-France*, Paris: INRETS.

Salomon, I., Bovy, P.H. and Orfeuil, J.P. (1993), *A Billion Trips a Day: Tradition and Transition in European Mobility Patterns*, Dordrecht: Kluwer.

PART IV

Modelling Behavioural Responses

10. Stated preference and conjoint analysis: A comparison using mode-choice behaviour

Kay W. Axhausen, H. Köll and M. Bader

1. INTRODUCTION

In spite of their known limitations and problems, stated-preference-based survey techniques have become an accepted part of the transport planning tool kit. In general they are implemented as stated-choice experiments, in which respondents are asked to choose between two or more alternatives described to them. The analysis of stated-preference data is decompositional in the sense that one derives the part-worths of the different variables describing the alternatives from the one, joint judgement of all attributes (A chosen over B; A ranked 5th, B 10th; A received a scale value of 6 out of 10). In the case of choice data, logit or related utility-maximizing models are used to derive those part-worths.

In marketing, surveys of hypothetical markets/goods are called conjoint analyses. The approaches used, for example the very popular ACA Software (Adaptive Conjoint Analysis) (Sawtooth, 1996), are hybrids of compositional and decompositional analysis methods. Compositional utility estimation is performed in two steps. In the first step, variables are ranked one at a time for their importance and their attribute levels are ranked for their desirability. The total utility of an alternative (composite good) is calculated as the importance-weighted sum of the desirabilities.

Conjoint analysis (CA) and stated preferences (SP) belong therefore to the same general class of techniques, where the respondents are offered hypothetical goods, mostly in the form of written descriptions, either individually or in sets, which they are asked to rate, rank, or choose between (Hensher, 1994; Axhausen, 1996). Both approaches are based on earlier work in psychology and microeconomics (Green and Rao, 1971; Louviere et al., 1973). Users of both techniques accept that the true complexity of the decision studied is higher than the level presented to the respondent in the survey. This simplification is accepted by the users as the price to obtain results of predictive value for those factors under the control of the authority or firm undertaking the study within an acceptable time frame (see Brög, 1997, for an opposing view).

The stated-preference and the hybrid conjoint-analysis approaches encompass a wide variety of specific methodologies, which all share the aim of obtaining holistic statements of preference in a specified format for a series of (hypothetical) goods described by varying levels of a small number of attributes according to a well-designed experimental plan within a specified behavioural frame (overall context).

The development of conjoint analysis is well documented in a series of scholarly reviews in the marketing research literature (Green and Srinivasan, 1978; Böcker, 1986; Huber, 1997; Louviere, 1988). The growing usage of the methodology in marketing over the last two decades is well documented by three surveys of market research firms (Cattin and Wittink, 1982; Wittink and Cattin, 1989 and Wittink et al.,1994). The development of the methodology in transport planning can be reconstructed from a series of how-to manuals, which have been published over the years (Kocur et al., 1982; Pearmain et al., 1991; Axhausen, 1996; Pearmain et al., forthcoming; but see also Bates, 1988 or Hensher, 1994). The usage of the methodology in transport planning has not been reviewed yet.

The current return of SP to the USA, the coming together of professional market researchers and transport planners in the commercialized public transport firms (bus, rail and air) and the growing interest in market research in choice-based conjoint formats (Louviere and Woodworth, 1983; Sawtooth, 1995) opens up new opportunities for the further development of both methods.

The purpose of this chapter is to compare two particular SP and CA methods with regard to the comparability of their results using mode-choice behaviour in the City of Innsbruck to gain an insight into the potential of each of the approaches as a policy research tool.

The remainder of the chapter is structured as follows. The next section describes the two approaches implemented in the survey, which are then described in detail. The survey administration and the response behaviour are analysed in the following section, while the results from the modelling of the responses is the topic of the final section. A summary and a discussion of further work concludes the chapter.

2. SURVEY APPROACH

The methodological interest had to be balanced with the substantive interests of all parties involved in funding the survey, which required the drawing of a large sample of respondents. This requirement led to the choice of a combination of a telephone survey with a follow-up postal survey, which was based on the answers in the telephone survey (for the pioneering study see Polak et al., 1991). Ideally one would have wished to use a computer-based interview for the CA/SP elements of the work.

The telephone survey covered the following topics:

- availability of public transport at home and at work, where relevant, in terms of distance to the nearest stop and number of lines available;
- availability of a car or of a public transit pass;
- availability of parking at home and at work, in terms of distance to the parking space, its type and its costs;
- socio-demographic description of the respondent, including the ownership of a driving licence;
- a recent trip to either work, shopping or an evening leisure activity within the City of Innsbruck, including destination, access, wait, in-vehicle, parking search and egress times, transfers, availability of seat, means of public transport (bus, trolley or tram), fare and parking fee (for the chosen and the competing modes); and
- the number of trips undertaken by public transport during the past week and the usage of different ticket types (one-half of the sample reported numbers and usage for the week as a whole, while the other half reported trips per day and the ticket used on that day for each of the seven days).

The information was coded and the trip descriptions were used to generate postal SP and CA surveys, which were sent to the respondents within generally four (two working) days. See Table 10.1 for the variables obtained in the telephone survey, which were systematically varied for the SP and CA studies.

Conjoint-analysis Survey

CA covers a whole range of different approaches, which calculate individual utility part-worths using both compositional and decompositional approaches (Schubert, 1991). The chosen hybrid approach combines both compositional and decompositional elements (Green and Krieger, 1996) by offering first a series of rating tasks, in which the respondent has to judge the importance of an attribute and the desirability of different levels of the attribute, and by then offering a set of alternatives, described with all relevant attributes, which the respondent has to rate as a whole. The first part allows the estimation of utility by adding (composing) it from the assessment of individual attributes and their levels. The second part allows the estimation of the part-worths of

Table 10.1 Variables used in the CA/SP tasks

Attribute (number of levels)			
Public transport	Car	Bicycle	Walking
Means of transport (3)			
Access time (2)	Access time[b] (–)	Access time[a] (2)	
Headway (2)			
Waiting time[a] (2)			
Transfer (2)			
In-vehicle (inclusive of transfer times) (2)	In-vehicle (without search) (2)	Riding time (2)	Walking time (2)
	Parking search time (2)	Parking search time[b] (–)	
Egress time[b] (–)	Egress time[a] (2)	Egress time (2)	
	Type of parking (3)	Type of parking (2)	
Reliability (probability of lateness) (2)	Reliability (probability of congestion) (2)		
Fare (3)	Parking fee (2)		
		Share of bicycle paths (3)	

Notes:
[a] Only for the CA compositional tasks.
[b] Not varied for the CA full profiles/SP tasks.

the different attributes by decomposing the joint rating of the alternative offered. (See Figures 10.1, 10.2 and Appendix 1 for a translation.)

Each respondent was sent 14 tasks of the first type (five pages with three each, including the reported mode from the reported target trip) (see Figure 10.1 for an example) and 14 tasks of the second type (five pages with three each, including the reported mode) (see Figure 10.2 for an example). The attribute values were varied consistently around those reported for the target journey. Depending on the availability of a car to the person, the modes presented were public transport and car or public transport, bicycle and walk.

Table 10.1 lists the attributes used. The experimental design was a random sample of the 2^{11} 3^3 full factorial (44 situations, which were divided into four blocks with some overlap). The sample of situations was checked for the extent of the correlation between the attributes and the absence of factorial structures. The tasks related to the recorded work and shopping trips.

Wie wichtig ist für Sie bei der Fahrt zur Arbeit mit einem öffentlichen Verkehrsmittel der Fahrpreis?

ganz unwichtig										sehr wichtig
0	1	2	3	4	5	6	7	8	9	10

Wie beurteilen Sie folgende Preise für die Strecke, die Sie uns geschildert haben?

	unattraktiv										attraktiv
17 Schilling											
20 Schilling											
24 Schilling											
	0	1	2	3	4	5	6	7	8	9	10

Figure 10.1 Conjoint-analysis survey: example of attribute and level rating task

Angenommen, es gäbe die folgende Möglichkeit:

Auto:	Fahrzeit mit dem Auto ist	13 min
	Stau von 5 Minuten oder länger	an 3 von 10 Tagen
	Parkplatzsuche dauert gewöhnlich...........	8 min
	Parkmöglichkeit am Ziel	am Straβenrand
	Parkplatz kostet	48 Schilling pro Tag

Ihre Bewertung wäre:

unattraktiv										attraktiv
0	1	2	3	4	5	6	7	8	9	10

Figure 10.2 Conjoint-analysis survey: example of a full-profile rating task

Stated Preference

The stated-preference element of the survey was implemented as a stated-choice experiment with respondents choosing between car, public transport, bicycle and walking, if a car was available, and public transport, bicycle and walking, if no car was available. In the first case, bicycling and walking are described as 'as is', without giving any details of attribute levels. The other two modes were varied systematically. In the second case, the descriptions of all three modes were varied. In the case of public transport the access and egress walking times are presented as their sum, while the in-vehicle times include any transfer times. The in-vehicle time for the car excludes any parking search time. Access times to the car are assumed to be constant at current values.

The experiments were conducted for three trip purposes (work, shopping and evening leisure). Each respondent received 11 choice tasks, plus a description of the reference journey reported in the telephone interview (six pages with two descriptions each) (see Figure 10.3 for an example and Appendix 1 for a translation).

Angenommen, die Situation wäre nun so:		
Öffentlicher: Verkehr:	Es fährt ein	Bus
	Bus fährt	alle 20 min
	Bus ist	in 0 von 10 Fällen unpünktlich
	Umsteigen	ja
	Fahrt mit Umsteigen dauert	insgesamt 23 min
	Fuβwege von/zur Haltestelle dauern	insgesamt 14 min
	Fahrt mit dem Bus kostet	20 Schilling
Auto:	Fahrzeit mit dem Auto ist	8 min (ohne Parkplatzsuche)
	Stau von 5 Minuten oder länger........	an 0 von 10 Tagen
	Parkplatzsuche dauert gewöhnlich	7 min
	Parkmöglichkeit am Zeil	am Straβenrand
	Parkplatz kostet	36 Schilling
Rad:	Wie derzeit	
zu Fuβ:	Wie derzeit	
Ihre Entscheldung wäre:	Bus .... ☐ Auto ... ☐ Rad ... ☐ zu Fuβ ... ☐	

Figure 10.3 SP experiments: example of a choice task

3. SURVEY ADMINISTRATION AND RESPONSE BEHAVIOUR

Survey Administration

The survey work was conducted in two parts during the winter of 1997 (November/December 1997 and February/March 1998) to avoid the clash with the Christmas holidays and the local school holidays at the beginning of February.

The sample addresses of households in the City of Innsbruck were obtained from an address dealer (addresses and current telephone numbers). The numbers were screened against the current post office compact disc of telephone numbers and any erroneous addresses were discarded. Every address/telephone number was tried five times at different times of the day over a number of days before it was classified as unreachable. To obtain a random sample of persons, we asked to speak to the person with the birthday closest to the date of the interview. The type of target trip purpose was allocated randomly to the person, but so as to maintain a balance between the trip purposes. The same applies to the allocation between the SP and CA experiments.

The forms for the CA/SP experiments were sent in general within four (two working) days after the telephone interview. A reminder call was made if no response was obtained within two weeks of sending the survey. The average respondent took ten days to return the forms.

Response Behaviour

Table 10.2 summarizes the overall response behaviour. The share of unreachables is typical for the City of Innsbruck, reflecting the substantial share of second homes in the City. The share of those reached who completed the interview was satisfactory at 66 per cent, of which nearly all had a suitable target trip to report.

The response rate for the SP/CA experiments was identical: a satisfactory 65 per cent. The response behaviour was analysed using probit models of the

Table 10.2 Response behaviour

Response							Share of all
Unreachable	391			(18%)			18%
Reached	1832			(82%)			
Refused		487			(27%)		22%
Aborted		130			(7%)		6%
Full interview		1215			(66%)		
With trip			1161			(96%)	52%
Without trip			54			(4%)	2%
Sum	2223	1832	1215				2223

response probability using the available set of socio-demographic variables contrasting those who had participated in the telephone interview but had not returned the forms with those who did. The equations estimated were not significant overall and only a small set of variables had a significant impact, but there was no overlap between those significant in the CA response model and those significant in the SP response model. The willingness to participate in such survey tasks seems therefore unrelated to the socio-demographic characteristics of the respondents. The commitment comes from other sources, which cannot be described with the socio-demographic variables available here.

The telephone interview technique led to an overrepresentation of older and female respondents. The sample was therefore weighted to reproduce the known distribution of residents with regard to age (three age categories), sex and public transit season ticket ownership.

4. RESULTS FROM THE CA/SP EXERCISES

Analysis of the Hybrid CA

The hybrid approach chosen here requires that the compositional and the decompositional elements of the exercise be brought together in one uniform analysis framework. Adapting the procedure suggested by Green and Krieger (1996), the following algorithm was implemented:

1. Calculate the ratings y_{ijk} for each level i of each attribute j for each person k from the compositional questions as:

$$y_{ijk} = d_{ijk} w_{jk}$$

with

$$\sum_{\forall j} w_{jk} = 1.0$$

using the desirability ratings d_{ijk} of the levels and the scaled importance rating w_{jk} of each attribute.

2. Calculate the scaled ratings y_{nk} for each full profile n for each person k from the decompositional tasks as:

$$y_{nk}^{i+1} = \frac{y_{nk}^{i} - \mu}{\tau}$$

using the rating r_{nk} as y_{nk}^{0}. In the first iteration assume the scaling parameters μ and τ to be zero and one.

3. Construct a joint data matrix from steps 1 and 2 as:

$$\begin{bmatrix} y_{ijk} \\ y_{nk} \end{bmatrix} = \begin{bmatrix} 0 \\ 1 \end{bmatrix} + \begin{bmatrix} V_{jk} \\ V_{rk} \end{bmatrix} 2$$

with the vector *y*s of the ratings and a vector 0 of zeros, a vector 1 of ones and the vector *V* describing the values of the levels of the attributes. V_{jk} consists of zeros for the non-rated attributes *j* and of the rated value of the level of the attributed rated. V_{nk} consists of the values of levels of the attributes X_{jnk} in the full profile.

4. Estimate with multiple linear regression the ßs for the attributes:

$$\hat{y} = \beta_j X_j + \varepsilon$$

5. Re-estimate μ and τ as the intercept and slope of the simple linear regression of the model:

$$y_{nk}^{i} = \mu + \tau \hat{y}_{nk}^{i} + \varepsilon$$

6. Repeat steps 2 to 4 until the sum of the squares of the errors of the regression in step 3 changes less then a predetermined amount between iterations.

This procedure, which essentially scales the ratings from the decompositional tasks to the mean and variance of the ratings from the compositional task, converged well in this application (three to five iterations with a stopping criterion of 3 per cent change between iterations). The calculations were performed with the linear regression procedure of Statistical Analysis System.

Analysis of the SP

The data from the SP were analysed using the procedure NLOGIT of LIMDEP 7.0 (Econometric Software, 1998). Persons who chose one mode only across all 11 choice tasks were removed from the estimation of the multinominal logit models reported below. The travel times of the 'as is' condition for the cyclists and pedestrians were estimated from the zone-to-zone car travel times of an available assignment model for the City of Innsbruck, which were scaled using the reported travel times for these modes. The SP data were merged with the available RP data, that is, the trips reported, which formed the basis of the SP exercise, to improve the quality of the estimates. The required tree structure placed the RP choices in one nest, while the SP-based alternatives were allocated to four different nests, one for each alternative (see Bradley and Daly, 1993). The SP data were re-weighted, so that the RP data and all SP data together had equal weight in the analysis. The method scales the error variances of the SP alternatives relative to the error variances of the RP alternatives, while assuming that the parameters are the same in the SP and RP data set. The scaling parameter λ indicates the strength of the scaling required.

Results

To gain an initial understanding of the data, models were estimated for both the CA and the SP/RP data employing the relevant variables with linear terms only and including the available socio-demographic variables for each person (sex, age in decades (set of dummy variables), public transport season ticket ownership, employment status, student status, recipient of a high-school diploma). More complex forms (logistic transformation of the desirabilities for the CA or quadratic terms of the independent variables) did not increase the explanatory value of the models and will therefore not be reported here. The CA models were estimated separately for each mode and trip purpose.

The significance levels of the parameters were corrected by either the square root of the number of cases or the third root of the number of cases per person to account for the repeated-measures problem in both the CA and the SP exercises (Bates and Terzis, 1997). In the tables below the first correction (columns marked 1/2) is deemed in general to be too conservative, while the second correction (columns marked 1/3) is deemed to be more appropriate in the absence of a more rigorous estimation procedure (for example models allowing for taste variation or serial correlation between answers).

The results of the analysis of the CA data are presented in Table 10.3 for the car attributes, in Table 10.4 for the public transport attributes and in Table 10.5 for the bicycle attributes. A number of results are not consistent with expectations, as for example in the case of the car, the positive parameters for the walking and driving time, which are even in some cases significant. Equally surprising are the extremely high values for transferring and reliability in the case of public transport, which seem unrealistic.

Initial analyses of the SP and RP data not reported here showed that the parameter estimates are in many cases rather similar across the modes. Therefore a joint SP/RP model using generic parameters where possible was estimated focusing on the modal parameters and using the same set of socio-demographic variables as before.

The estimation of the impacts of reliability proved problematic due to the RP data, which were unsatisfactory in their description of reliability (too

Table 10.3 CA: results for the car attributes by purpose

Attribute		Purpose								
		Work Parameter	1/2 sig.	1/3 sig.	Shopping Parameter	1/2 sig.	1/3 sig.	Leisure Parameter	1/2 sig.	1/3 sig.
Walking time	[min]	0.108			0.213	*	*			
Travel time	[min]	0.065		*	0.022					
Search time	[min]	–0.198		*	–0.181	*	*		Not	
Reliability	[]	–0.166			–0.058	*	*		available	
Parking fee	[S]	–0.042	*	*	–0.229		*		for CA	
Parking lot	[y/n]	0.736			0.762					
Garage	[y/n]	0.892			0.484					
F		25.109			34.735					
adj. R^2		0.174			0.226					
N		2066			2075					
VOT	[S/min]	–1.548			–3.672					
Walking/travel	[]	1.662			9.682					
Search/travel	[]	–3.046			–8.227					
Reliability/travel	[]	–0.646			–10.409					

Table 10.4 CA: results for the public transport attributes by purpose

		Purpose								
		Work			Shopping			Leisure		
Attribute		Parameter	1/2 sig.	1/3 sig.	Parameter	1/2 sig.	1/3 sig.	Parameter	1/2 sig.	1/3 sig.
Walking time	[min]	–0.064			–0.071		*			
Travel time	[min]	–0.034		*	–0.009					
Headway	[min]	–0.027			–0.023			Not available for CA		
Reliability	[]	–0.271		*	–0.233		*			
Fare	[S]	–0.004			–0.031					
Transfer	[y/n]	–2.341	*	*	–1.857	*	*			
Trolley bus	[y/n]	–0.301			0.043					
Street car	[y/n]	–0.121			–0.033					
F		24.222			29.750					
adj. R^2		0.125			0.135					
N		3242			3487					
VOT	[S/min]	8.500			0.287					
Walking/travel	[]	1.882			7.978					
Wait/travel	[]	1.588			5.169					
Transfer/travel	[]	68.853			208.652					
Reliability/travel	[]	7.971			26.180					

Table 10.5 CA: results for the bicycle attributes by purpose

Attribute		Purpose								
		Work Parameter	1/2 sig.	1/3 sig.	Shopping Parameter	1/2 sig.	1/3 sig.	Leisure Parameter	1/2 sig.	1/3 sig.
Walking time	[min]	–0.258			0.042					
Travel time	[min]	–0.020			0.023					
Search time	[min]	1.511			–0.407			Not available for CA		
Cycle paths	[%]	0.019			0.034	*	*			
Cycle stand	[y/n]	2.580	*	*	1.686		*			
F		13.356			15.331					
adj. R^2		0.305			0.263					
N		450			521					
Walking/travel	[]	12.900			1.826					
Search/travel	[]	–75.550			–17.696					
Cycle path/travel	[]	–0.950			1.478					
Stand/travel	[]	129.000			73.304					

Table 10.6 SP/RP: results for the modal attributes

Attribute		Purpose								
		Work Parameter	1/2 sig.	1/3 sig.	Shopping Parameter	1/2 sig.	1/3 sig.	Leisure Parameter	1/2 sig.	1/3 sig.
Walking time	[min]	–0.090	*	*	–0.070	*	*	–0.101	*	*
Travel time	[min]	–0.054	*	*	–0.062	*	*	–0.032	*	*
Headway	[min]	–0.015		*	–0.004			–0.007		
Search time (car)	[min]	–0.091	*	*	–0.197	*	*	–0.106	*	*
Search time (cycle)	[min]	–0.111			–0.307			–0.931		*
Transfer	[y.n]	–0.775	*	*	–0.562	*	*	–0.711	*	*
Fare/Fee	[Schilling]	–0.021	*	*	–0.026	*	*	–0.048	*	*
Reliability (PT)	[x/10]	**–0.540**			**–0.360**			**–0.270**		
Reliability (car)		**–0.400**								
Cycle paths	[%]	–0.004			0.013	*	*	–0.005		
Trolley bus	[y,n]	0.785	*	*	0.534	*	*	–0.257		
Street car	[y,n]	–0.316			0.560	*	*	0.730	*	*
Parking lot	[y,.n]	0.572	*	*	2.533	*	*	–0.005		
Multi–storey	[y,n]	0.609	*	*	1.256	*	*	0.822		*
Cycle rail	[y,n]	0.213			–0.596			0.648		

Summary statistics									
λ	0.956	*	*	0.312	*	*	0.384	*	*
$\rho^2(0)$	0.725			0.759			0.684		
$\rho^2(C)$	0.279			0.223			0.162		
N	1095			1692			1312		

Notes:
The parameters in bold were fixed based on the earlier SP results.
The significance levels are based on 1.96*square root(2) (=1/2 sig) and 1.96*third root(2) (1/3 sig.), as there are two observations per person given the weighting of the SP observations.

$\rho^2(0) = L(\beta) - L(0) / L(0)$; $\rho^2(C) = L(\beta) - L(C) / L(C)$;

little variance, a difficult-to-understand description of unreliability in the case of car transport). The parameters were therefore fixed a priori using the earlier SP results.

Table 10.6 presents the results for the modal attributes and the summary statistics. The fits of the models are good. The highly significant scaling parameters λ indicate the necessity of the estimation method used.

The core modal attributes are all significant, with the exception of headway/waiting time, which is only (weakly) significant in the case of work. The introduction of the reliability variable must be the reason for this, as the headway variable normally captures both aspects of the wait (waiting time and the unreliability of the service). The transfer variable captures the inconvenience of the transfer plus the chance for additional unreliability. The importance of cycle paths to cycling is visible in the joint estimation.

The relative valuations and the values of time are shown in Table 10.7. The values of travel time savings are in a realistic range of about 2.0–2.5 Schilling/min for work and shopping, while the value is considerably lower, as expected, for leisure. The relative valuations for walking show a reasonable pattern, although the value for shopping seems low in comparison with many other studies. The separate estimation of waiting time and unreliability results in comparatively low valuations for waiting time, which are in turn balanced by substantial valuations for unreliability, especially for the work trip.

The search time valuations for car travel are in the expected range, while the valuations for cycling seem high, but have to be accepted in the absence of comparable results as the expression of the ardent wish to park one's bicycle in front of the destination, in particular during shopping and at night. The transfer penalty of about 10–15 minutes during the day (work and shopping) is within the range of earlier studies, while the value for leisure (22 minutes) seems reasonable given the associated loss of comfort, especially during the evening and night.

The results for the different types of public transport vehicle are more conclusive for the joint estimation, indicating, in comparison with the normal bus, a preference for the trolley bus to work, the trolley bus and the street car to shopping, and for the street car to a leisure activity. Similarly, there are significant preferences for off-street parking during work and shopping.

5. CONCLUSIONS AND FUTURE WORK

The CA models produce fewer significant parameter estimates and more estimates that seem unrealistic in comparison with prior knowledge. The signs of the estimates are in general the same and the rank order of the

Table 10.7 SP/RP: relative valuations

	Purpose											
	Work				Shopping				Leisure			
Relation	All modes	PT	Car	Bike	All modes	PT	Car	Bike	All modes	PT	Car	Bike
VOT	2.57				2.38				0.67			
Walking/travel	1.67				1.13				3.16			
Waiting/travel	0.56				*0.13*				*0.44*			
Searching/travel	1.69	2.06			3.18	4.95			3.31			29.09
Transfer/travel	14.35				9.06				22.22			
Reliability/travel	**10.00**	**7.41**			**5.81**				**8.44**			
Cycle path/travel	*–0.07*				*–0.21*				*0.16*			

Notes:
Bold values were determined a priori.
Italicized values are based on non-significant parameter estimates.

weights is also normally identical, but the relative sizes can vary considerably, raising doubts about the consistency of either set of results.

CA estimates of the value of time for in-vehicle time are low, but not unreasonable. It is interesting to note that for work the estimates for the public transport fare are not significantly different from zero, reflecting on the one hand the long-term commitment of a public transport season ticket holder and on the other hand the necessity to use public transport for the other users. The parking fee estimates are consistently significant for work (CA, SP and SP/RP models).

The relative valuations for the different time elements vary considerably, but for the SP and RP case they do not deviate massively from prior expectations, except for walking time relative to travel time, which seems low reflecting the lack of variability in the data. The CA estimates are in a fair number of cases excessive.

It is difficult to judge to what extent these unexpected patterns are due to the presence of the reliability variable, which does not produce convincing results other than in the SP context. For the RP models (not reported here) it has the wrong sign and is significant twice, it is insignificant twice. This might be due to the lack of range and variability in the rather non-congested Innsbruck. For the CA and SP car models the estimates are either not significant or only marginally so, again reflecting either too little variability in the data or an unclear description of the variable, which might have been misunderstood by the respondents. (Some respondents might have included the congested time with the travel time specified for the CA/SP description.) The reliability estimates for the SP public transport models are significant and have the right signs (less significant for the CA models), reflecting an easier-to-understand formulation of reliability (x of out 10 late by 5 minutes) and more variability in the observed data.

The initial analyses indicate that the two approaches produce results consistent in their trends, but not necessarily in their exact valuations. Further work is needed to identify the reasons for those differences. The future work planned will address these challenges. In particular, it is planned to cross-validate the CA rating-based results against the SP choice-based results by building a choice simulator, which uses the utility part-worth estimates to predict choices for the SP choice tasks. Consistency at this level would be useful, even if consistency at the relative parameter estimates cannot be established.

A second important direction is the estimation of individual parameter estimates from the CA exercise, which should shed new light on the distribution of the valuations of the modal attributes, in particular of reliability, of waiting time, and of the transfer penalty.

ACKNOWLEDGEMENTS

The survey work reported in this chapter was funded by the EU funded 4th Framework project TASTe, the Innsbrucker Verkehrsbetriebe and the City of Innsbruck. The authors are grateful for the comments received during the discussions at the 4th NECTAR conference and at the 26th European Transport Conference, where earlier versions of this chapter were presented (Axhausen et al., 1998a and b). The responsibility for the results and their interpretation rests with the authors alone.

REFERENCES

Axhausen, K.W. (1996), *Hinweise zur Messung von Präferenzstrukturen – die Methoden der "Stated Preferences"* (Stated Preference Guidelines), with contributions from W. Bogner, M. Herry, H. Verron, H. Volkmar and W. Wichmann, Köln: Forschungsgesellschaft für das Straßen- und Verkehrswesen.

Axhausen, K.W., H. Köll and M. Bader (1998a), 'Stated Preferences and Conjoint Analysis: A Comparison Using Mode Choice', paper presented at the 4th NECTAR Conference, Kibbutz Shefajim, April.

Axhausen, K.W., H. Köll and M. Bader (1998b), 'Experiments with SP and CA approaches to Mode Choice', Paper presented at the 26th ETC Conference, Loughborough, September.

Bates, J.J. (ed.) (1988), 'Stated Preference Methods in Transport Research', *Journal of Transport Economics and Policy*, **22**(1).

Bates, J.J. and G. Terzis (1997), 'Stated Preference and the "Ecological Fallacy"', *Proceedings of the 25th European Transport Forum*, **F2**, London: PTRC, pp. 155–70.

Böcker, F. (1986), 'Präferenzforschung als Mittel marktorientierter Unternehmensführung', *Zeitschrift für betriebswissenschaftliche Forschung*, **38** (7/8), 543–74.

Bradley, M.A. and A.J. Daly (1993), 'New Issues in Stated Preferences Research', presentation at the 21st PTRC Summer Annual Meeting, Manchester, September.

Brög, W. (1997), 'Raising the Standard: Transport Survey Quality and Innovation', paper presented at the International Conference on Transport Surveys: Raising the Standard, Grainau, May.

Cattin, P. and D.R. Wittink (1982), 'Commercial Use of Conjoint Analysis: A Survey', *Journal of Marketing Research*, **16**(1), 44–53.

Econometric Software (1998), *LIMDEP 7.0 for Windows*, Sydney: Econometric Software.

Green, P.E. and A.M. Krieger (1996), 'Individualized Hybrid Models for Conjoint Analysis', *Management Science*, **42**(6), 850–67.

Green, P.E. and V.R. Rao (1971), 'Conjoint Measurement for Quantifying Judgmental Data', *Journal of Marketing Research*, **8**(8), 355–63.

Green, P.E. and V. Srinivasan (1978), 'Conjoint Analysis in Consumer Research: Issues and Outlook', *Journal of Consumer Research*, **5** (September), 103–23.

Hensher, D.A. (1994), 'Stated Preference Analysis of Travel Choices: The State of Practice', *Transportation*, **21**, 107–33.

Huber, J. (1997), 'Conjoint Analysis: How We Got Here and Where We Are', paper presented at the Sawtooth Sotfware Conference, Evanston: Sawtooth Software Inc. Obtained as http://www.sawtoothsoftware.com/TechPap.htm/HowweW.zip

Kocur, G., T. Adler, W. Hyman and B. Aunet (1982), *Guide to Forecasting Travel Behavior with Direct Utility Assessment*, Report UMTA-NH-11-0001-82-1, US Department of Transportation, Urban Mass Transit Administration.

Louviere, J.J. (1988), 'Analyzing Decision Making: Metric Conjoint Analysis', *Sage University Papers*, **67**, Newbury Park: Sage.

Louviere, J.J., R.J. Meyer, F. Stetzer and L.L. Beaver (1973), 'Theory, Methodology and Findings in Mode Choice Behaviour', Working Paper, **11**, Institute of Urban and Regional Research, University of Iowa, Iowa City.

Louviere, J.J. and G. Woodworth (1983), 'Design and Analysis of Simulated Consumer Choice and Allocation Experiments: An Approach Based on Aggregate Data', *Journal of Marketing Research*, **20**, 350–67.

Pearmain, D., J. Swanson and E. Ampt (forthcoming), *Stated Preference Explained*, London: SDG.

Pearmain, D., J. Swanson, E. Kroes and M. Bradley (1991), *Stated Preference Techniques: A Guide to Practice*, London: Steer Davies and Gleave.

Polak, J.W., P.M. Jones, P. Vythoulkas, S. Meland and T. Tretvik (1991), 'The Trondheim Toll Ring: Results of a Stated Preference Study of Travellers' Response', Working Paper, **662**, Transport Studies Unit, University of Oxford, Oxford.

Sawtooth Inc. (1995), *The CBC System for choice-based Conjoint Analysis*, Evanston: Sawtooth Software Inc. Obtained as http://www.sawtoothsoftware.com/TechPap.htm/CBCTchWp. zip

Sawtooth Inc. (1996) *ACA System – Adaptive Conjoint Analysis – Version 4.0*, Evanston: Sawtooth Software Inc. Obtained as http://www.sawtoothsoftware.com/TechPap.htm/ACATchWp.zip

Schubert, B. (1991), *Entwicklung von Konzepten für Produktinnovationen mittels Conjointanalyse*, Stuttgart: Poeschel-Verlag.

Wittink, D.R. and P. Cattin (1989), 'Commercial Use of Conjoint Analysis: An Update', *Journal of Marketing*, **53** (7), 91–6.

Wittink, D.R., M. Vriens and W. Burhenne (1994), 'Commercial Use of Conjoint Analysis Europe: Results and Critical Reflections', *International Journal of Research in Marketing*, **11**(1), 41–52.

APPENDIX 1: TRANSLATION OF THE SP AND CA FORMS

Conjoint-analysis survey: example of attribute and level rating task (Figure 10.1)

How important is the public transport fare for you for your work trip?
Not important at all .... Very important
How do you rate the following fares for the trip you reported to us?
Unattractive .... Attractive

Conjoint-analysis survey: example of a full-profile rating task (Figure 10.2)

Assume the following were available:

Car: Driving time with the car .... 13 min
Congestion of 5 or more minutes .... 3 out of 10 days
Parking search normally takes .... 8 min
Parking at the destination .... Kerbside
Parking costs .... 48 Schillings per day

Your rating would be:
Unattractive .... Attractive

SP experiments: example of a choice task (Figure 10.3)

Assume the following situation:

Public transport
It is .... a bus service
The bus comes .... every 20 min
It is .... not on time in 0 out of 10 cases
Transfer .... yes
Riding time (inc. transfer) .... 23 min total
Access times .... 14 min total
Bus fare .... 20 Schilling
Car Driving time with the car .... 8 min (without parking search)
Congestion of 5 or more minutes .... 0 out of 10 days
Parking search normally takes .... 7 min
Parking at the destination .... Kerbside
Parking costs .... 36 Schillings per day
Bicycle As is
Walking As is

Your decision would be: Bus ...... ☐ Car ...... ☐ Cycle ...... ☐ Walk ...... ☐

11. Behavioural thresholds of commuters under congestion

Eliahu Stern

1. INTRODUCTION

Several studies have provided empirical evidence of the existence of thresholds in the decision-making process of commuters experiencing daily congestion. In 1979, Hutchinson and Dudeck (1979) found that drivers respond to varying lengths of expected delay in a positive and non-linear way. Later, Mahmassani et al. (1990) provided evidence of 'indifference bands', or thresholds, in commuters' travel choices. A more definitive result was presented by Khattak et al. (1993), who investigated delay thresholds by incorporating length of expected delay in their route diversion choice model. They found that drivers are more likely to divert if the expected delay time is at least 20 minutes. A mode-choice model developed by Koppelman et al. (1993) produced a set of threshold values representing boundaries of van pool intensity that affect modal switch intentions. Another quantitative indication of a threshold was found by Stern and Nasrin (1994). In this study, whenever there was an option to deviate from a congested road in the Tel Aviv metropolitan area, it was most likely that drivers would suffer a delay of at least 10 minutes before deviating. Later, Mahmassani and Rong-Chang (1998) developed specifications of indifference bands of schedule delay that govern both route and departure time choices. They showed that indifference bands are not constant over time, that older commuters are likely to tolerate greater schedule delay than younger ones, and that female commuters generally exhibit wider indifference bands than males. Finally, thresholds, either determined empirically or simply set to a pre-specified value, were used by Doherty and Axhausen (1998) in their model of the weekly household activity-travel scheduling process.

As more behavioural models are trying to incorporate thresholds in their behavioural mechanism, the present study is aimed at finding empirically the thresholds for two behavioural responses to congestion – activity behaviour and driving behaviour, within the context of decision field theory. Activity behaviour refers to the rescheduling and relocating of pre-planned work activities among suburban residents in the Tel Aviv metropolitan area. Driving behaviour examines the threshold for lane switching in terms of time pressure. The chapter is thus based on two surveys, each of which investigates a different response. It starts with a theoretical portion containing the conceptual framework of decision making under congestion and the dynamics of threshold change. Within this framework the results of the two surveys are presented and analysed. Finally implications and conclusions are drawn.

2. DECISION MAKING UNDER CONGESTION

Following the findings of Mahmassani and Rong-Chang (1996), the behavioural threshold is changing through time and among people. Under congestion it depends on both dynamic road situations and the socio-demographics of drivers. A choice theory incorporating such a dynamic threshold component, which aims to explain the high variability of human preferences in congested situations should have the following properties (Stern, 1998):

1. It should be of a process nature.
2. It should be of a risky type.
3. It should be descriptive.
4. It should suit dynamic situations.

The newly developed decision field theory combines all these properties. It incorporates, among other components, a threshold which governs the implementation of a momentary preference state. Since the theory, in its transportation context, has been presented elsewhere (Stern, 1998), only the relevant conceptual framework will be presented here.

The decision field theory (DFT), developed by Busemeyer and Townsend (1993) and later elaborated by Stern (1998), is aimed at understanding the motivational and cognitive mechanisms that guide a deliberation process involved in making decisions under uncertainty. It integrates both information from the external environment and information from the individual's associative memory as determinants of possible actions to be evaluated in a deliberation process. Deliberation is a time-consuming cognitive process that involves information seeking about the consequences (that is, events, E_i) of recognizable actions (A_i), weighting of the pay-offs (Y_i) of the events, and comparing the weighted consequences to establish a preference (choice) state. The deliberation process is manifested by vacillation until a choice is made and action is taken (Figure 11.1).

The theory introduces time elements (hA and hB) representing the amount of time that is taken to retrieve and process one pair of anticipated consequences

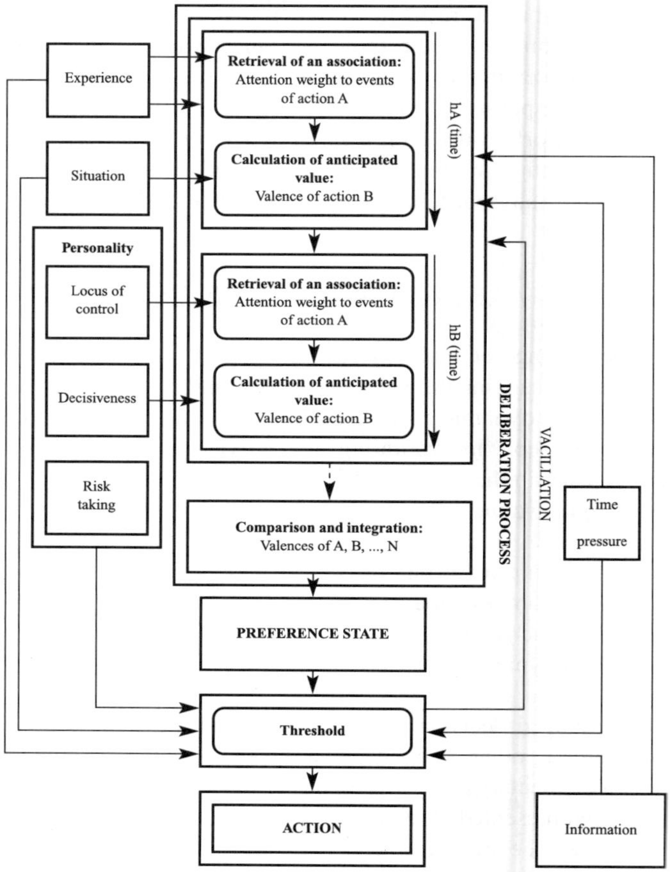

Figure 11.1 A DFT-based framework of choice behaviour

of actions (A and B) before shifting attention to another pair of consequences (of another action). The anticipated consequences are retrieved from associative memory. The total decision (deliberation) time equals the total time required to retrieve, compare and integrate all the comparisons of the anticipated consequences. The calculations involved in such a procedure are assumed to be realized by an underlying neural system and the DFT is an abstract representation of its essential dynamic properties.

The theoretical framework shown in Figure 11.1 presents a system in which exogenous variables (such as personality attributes, experience, situational characteristics and information) affect either the deliberation process and/or a threshold. The threshold is an inhibitory criterion presenting a given value of a momentary preference state. According to the decision field theory, the deliberation is a sequential process of calculating and comparing valences of *N* subjectively identified actions. The valence of an action at any moment represents the anticipated value of the action at that moment and it is produced by the weighted sum of the input values connected to that action. The time spent on the calculation of a given action (hA in Figure 11.1) depends on the attention given to the events of this action. The attention changes over time as a result of shifts in the decision maker's attention to the various consequences during deliberation. The attention given to the consequences of the events of alternative action B is not necessarily equal to that of action A and therefore, hA is not necessarily equal to hB. The result of one deliberation cycle is a momentary preference state which is chosen only when it reaches a threshold. Until then, vacillation is observed, thus activating another deliberation cycle. The smaller the distance between the preference state and the threshold, the smaller is the number of deliberation cycles.

According to this DFT conceptual framework, a given action is taken as soon as the momentary preference state exceeds a given value. The exogenous variables, including the situation shown in Figure 11.1 are assumed to affect the threshold. It is suggested that these effects can vary according to the way in which the threshold is defined. In general, the threshold is expressed by an inhibitory criterion which presents a given value of a preference state. Thus, a given response is taken only after a momentary preference state exceeds that given value. In transportation-related situations, the thresholds may have several different determinants and definitions, as argued below.

3. THE BEHAVIOURAL THRESHOLD

The empirical evidence cited in the introduction refers to delay time as the major determinant of the threshold. A driver stuck in a traffic jam, for example, may deviate only after a certain period of time. This length of time depends on the dynamic situation on the road and the driver's personality (as shown in Figure 11.1). Expected delay time, which usually depends on experience, acts in the same way as actual delay time only in cases where reactive behaviour (for example, deviating en route) is contemplated. When behaviour is preventive (for example, changing departure time due to expected congestion), there is a smaller effect of time pressure on the threshold (Stern, 1999).

Holding time pressure constant, levels of inherent aggression (a personality attribute) can inversely affect the value of the momentary preference state. As the level of inherent aggression increases, the inhibitory value decreases. Under increasing time pressure and/or under uncertain circumstances, a momentary aggression develops, adding further to the basic inverse effect of inherent

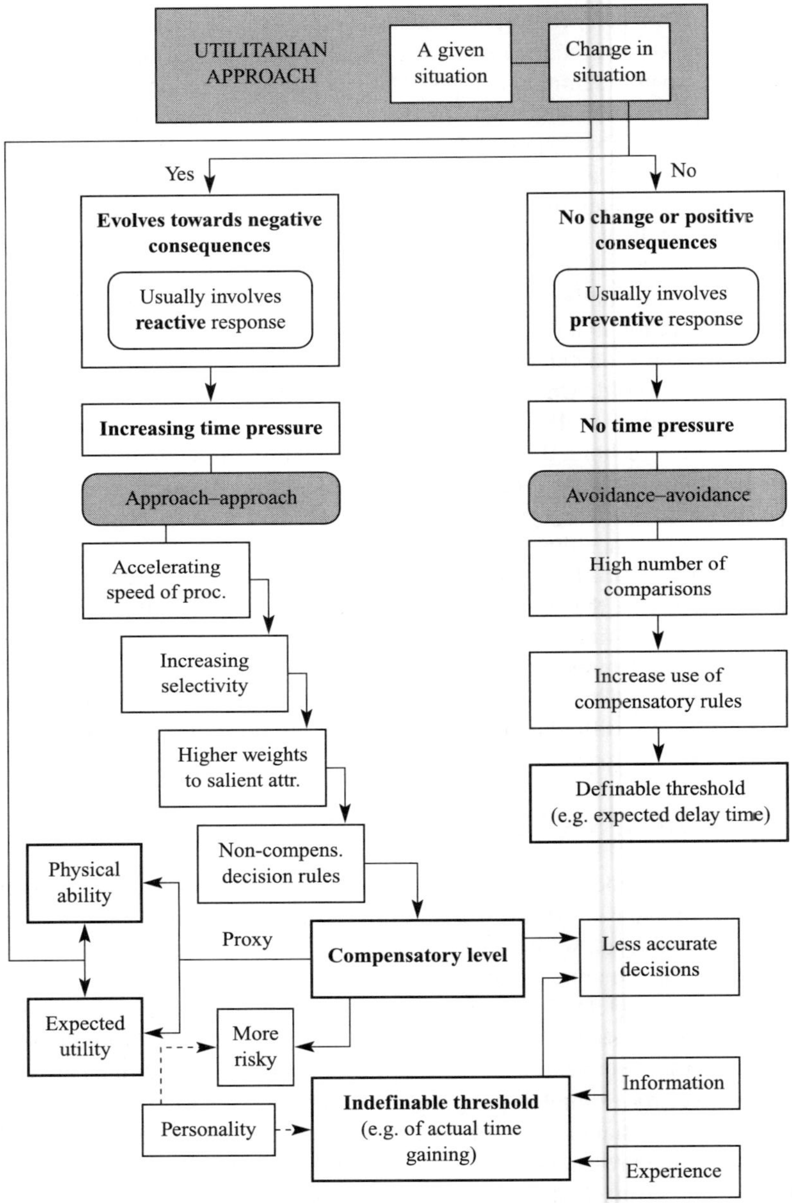

Figure 11.2 Situational change and threshold definition – a theoretical framework

aggression on the inhibitory value. The threshold could thus be expressed by aggression levels. The threshold can probably be expressed also by other determinants, such as monetary loss or amount of information, depending on the situation, personal tastes and attitudes. The delay time, however, is a threshold determinant which results indirectly from the DFT framework.

The DFT is a dynamic utilitarian theory which reacts to various situations. If the traffic situation does not change, or evolves toward positive consequences, individual responses are usually preventive. If the opposite occurs, and the situation evolves toward negative consequences, responses are usually reactive (Figure 11.2). The latter situation usually develops under increasing time pressure. These responses are explained by the approach–avoidance segment of the DFT (see Stern, 1998). Accordingly, the average weight given to a pay-off in a deliberation process is influenced by another dynamic variable called the 'goal gradient'. This variable reflects the relationship between the distance from one's goal and the weight given to the pay-off. The gradient of this relationship is changing whether the consequences of an action is a reward ('approach') or a punishment ('avoidance'). The basic idea is that the attractiveness of a reward or the aversiveness of a punishment is a decreasing function of the distance from the point of commitment to an action. Thus the consequences of an action become more salient as the preference state for that action approaches its threshold. The theory goes further, proposing that avoidance–avoidance (av–av) decisions produce longer mean deliberation times than do approach–approach (ap–ap) decisions. When av–av decisions are taken, drivers tend to make a high number of comparisons and the preference state thus vacillates and slows down the decision process. When ap–ap decisions are taken, drivers are more selective and the preference state is thus racing toward the threshold, shortening the decision time. This mechanism is further detailed in Figure 11.2.

In an av–av decision-making situation, there is no time pressure and individuals tend to use more compensatory decision rules. In such cases a threshold can be empirically defined, usually in terms of expected delay time. In an ap–ap decision-making situation, the degree of time pressure determines the form of the threshold. Since under ap–ap situations individuals tend to use more non-compensatory decision rules, a given compensatory level can only be used as a proxy to a threshold. The threshold is indefinable because it depends on various personality attributes, and other situational and personal variables (Figure 11.2), resulting in less accurate decisions and more risky behaviour. The proxy to the behavioural threshold depends on the dynamics of the situation. In a very dynamic situation, if the momentary preference state has not yet been defined, the threshold will be defined by the expected utility; if the momentary preference state has already been defined, the actual ability to respond is likely to become a physical threshold.

Based on two field surveys, each of which investigates a different response, the following sections will empirically define thresholds in both av–av and ap–ap situations. Activity behaviour presents an av–av situation in which thresholds for rescheduling and relocation of pre-planned work activities are defined, and driving behaviour presents an ap–ap situation in which thresholds for lane switching are defined.

4. THRESHOLDS OF ACTIVITY BEHAVIOUR

Definitions and Data Source

Activity behaviour is one of the preventive responses to congestion (see for example, Stern et al., 1998). Preventive response is adopted in cases of recurrent congestion as well as in the presence of pre-trip information referred to either recurrent or non-recurrent incidents. In such cases activity behaviour will have both spatial and temporal dimensions. The spatial dimension will usually refer to changing location of activities due to expected congestion, while the temporal dimension will refer to the schedule of individuals' work and non-work activities. Changes in work scheduling due to expected, or reported, congestion include both daily working times and working days, and chaining of daily activities. The same may be applied for non-work out-of-home activities where people change their schedules. Very few studies have investigated the effect of congestion on activity behaviour. Out of the few are the 'before' and 'after' studies undertaken by Tacken and De Boer (1989, 1991) focused on the way employees use flexible working hours to avoid peak-hour traffic.

The present study focuses on the general public, examining the rescheduling and relocation of pre-planned work and non-work activities among suburban residents in the Tel Aviv metropolitan area. Data on activity behaviour were collected from commuters interviewed at four gas stations located on the main commuting arterials crossing the metropolitan middle ring towards the metropolitan core – the city of Tel Aviv. The sample included 160 commuting drivers interviewed throughout the daily business hours (07:00 a.m. to 19:00 p.m.).

Patterns of Activity Behaviour

The high level of congestion in the Tel Aviv metropolitan area (Stern, 1996) affects all those commuting to the metropolitan core. Results from the survey reveal that 46 per cent of the commuters have at times changed their activity behaviour due to (reported) existing, or anticipated, congestion (Table 11.1).

Table 11.1 Characteristics of change in activity behaviour

Characteristics	Percentage distribution
Change of activity behaviour	
Yes	46
No	54
Type of activity	
Work-related	87
Shopping/entertainment-related	13
Frequency of change	
Once a day	13
At least once a week	38
At least once a month	29
Less often	20
Type of change	
Change of time	54.3
Change of location	24.3
Cancel activity	21.4
Importance of activity	
Very important–important	53
Quite important	26
Not important	21

The highest rate of such response has been at least once a week, and mainly (87 per cent) concerned with work-related trips. Over 75 per cent of the changes in activity behaviour related to activities which were subjectively perceived as important.

The distribution of responses by type of change shows that the most common (54.3 per cent) is a change in activity scheduling and next (24.3 per cent) is a change in activity location. Severe congestion has led to cancellation of activities in 21.4 per cent of the cases. Each of these response patterns has a different threshold as presented below.

Thresholds of Response

The subjective threshold for any change in activity behaviour in terms of expected delay time has a mode of 60 minutes and an average of 89 minutes, with a large standard deviation of 54 minutes. There is, however, a difference between females and males (Table 11.2). For 42 per cent of the males the threshold will be within the 31–60 minutes band while only 27 per cent of the

Table 11.2 Thresholds of activity behaviour by gender

Gender	Threshold band (in minutes)						
	1–30	31–60	61–90	91–120	121–150	151–180	181–210
Female	17	27	20	4	5	23	4
Male	12	42	9	12	7	6	2
Total	14	39	14	11	6	13	3

females will be within that band and 23 per cent of them will be within the 151–180 minutes band.

A significant difference in the response threshold was also found between the types of activities (Figure 11.3). Non-work trips are found to be more sensitive, reaction-wise, to congestion than work trips. The thresholds' mode for the former is within the 1–30 minutes band while the respective mode for work trips lies within the 31–60 minutes band. It is worth mentioning the finding that the thresholds' mode is within the same 31–60 minutes band regardless of the frequency of response.

Different thresholds were found for different responses. The average expected delay time which will trigger a change of: activity location is 75

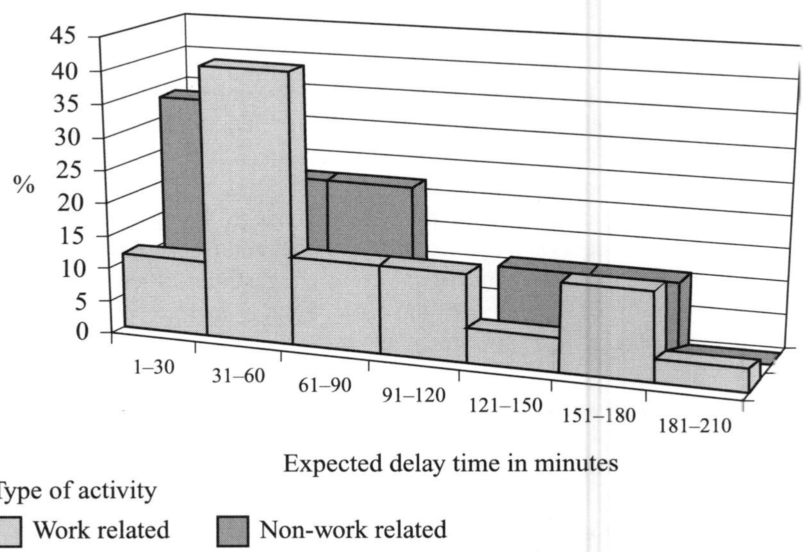

Figure 11.3 Thresholds of response by type of activity

minutes (standard deviation = 41 min.); activity scheduling is 86 minutes (S.D. = 49 min.); and cancellation of activity is 112 minutes (S.D. = 68 min.). Although a change in activity location was found to be the first expected response to congestion (that is, it has the smallest threshold), Table 11.1 shows that it was only second in practice, after activity rescheduling, presumably because it requires more demanding logistics than only changing time. Finally, an examination of thresholds by status at work (Figure 11.4) revealed a difference between self-employed and employees. While most employees, both regular workers and managers, will tolerate a 31–60 minutes delay before changing their activity behaviour, an equal number of self-employed commuters will tolerate either a lower threshold band (1–30 minutes) or a much higher (91–120 minutes) band. This certainly depends on the personal situation of the self-employed, both temporally and spatially.

In sum, it was found that the most common threshold band is 31–60 minutes, regardless of the type, frequency and place of response. Work trips were found more 'sensitive' to delays, with a more even distribution of the threshold bands. The activity behaviour threshold of self-employed commuters is significantly different from that of other employees, with the former having a wider band due to their higher freedom of choice. Finally, male commuters have a narrower threshold band than do female commuters. Much shorter time bands are expected to be found in ap–ap situations, as discussed below.

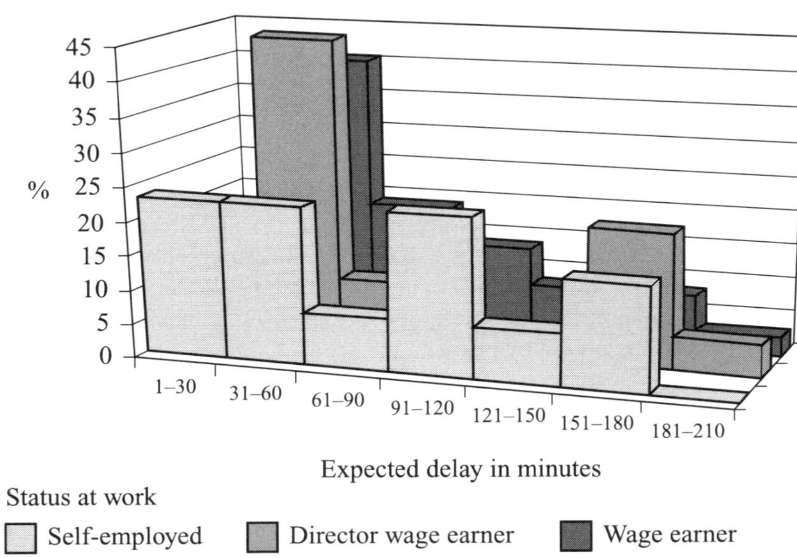

Figure 11.4 Thresholds of activity behaviour by status at work

5. THRESHOLDS OF DRIVING BEHAVIOUR

Thresholds' Definition

Driving behaviour is one of the reactive responses to congestion (for example Stern et al., 1998). Reactive response is adopted in non-recurrent incidents of congestion, or as a response to en-route information, usually under time pressure. It stems from the concept of impedance (Novaco et al., 1990), signifying a behavioural constraint on movement and goal attainment. It is claimed that as an aversive, frustrating condition, transportation impedance elevates arousal and elicits negative affect, thereby producing stress reactions that become relevant to both travel and driving behaviours. This concept was used by Adler et al. (1993) in calculating personal threshold tolerance for predicting en-route driver behaviour in response to real-time traffic condition information. The most observed driving responses to transportation impedance are by-law and aggressive lane switching, illegal on-kerb by-passing, illegal turns, purposeful deceleration (as a mean of 'punishment'), and use of prohibited lanes.

Among these driving responses to congestion, the present study focuses on lane switching. The threshold for lane switching changes according to the level of time-pressure and is expressed in terms of compensation level, as previously argued and presented in Figure 11.2. The compensation level is a proxy for expected utility (for example time gain) and can be only estimated by the difference in the number of drivers under low time-pressure who will definitely change lanes and the respective number of drivers who change under high time-pressure. As this difference increases, we assume that it should reflect the following trends: under low time-pressure (which is usually the case in av–av situations) the compensation level will increase because deliberation time increases due to the growing use of compensatory decision rules; and under high time-pressure (which is usually the case in ap–ap situations) the compensation level will decrease because deliberation time is getting shorter due to the use of non-compensatory decision rules. These hypothesized relationships are presented in Figure 11.5. An empirical test of these hypotheses is presented below.

Data Source

Threshold proxies for lane switching according to the hypotheses presented in Figure 11.5 were found through laboratory experimentation which was part of a larger study on the role of time-pressure in congested situations. The full details can be found in Stern (1999). In general, the study was based on experimentation which involved lane-switching decisions on a congested

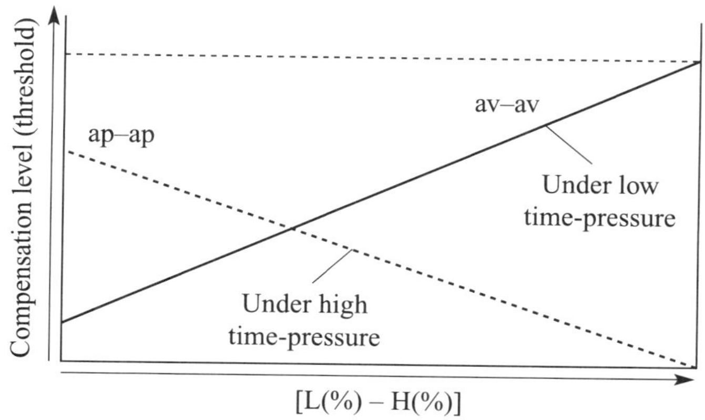

L – Percent of drivers under low time-pressure (the 25 sec. group) who definitely switch lanes.
H – Percent of drivers under high time-pressure (the 5 sec. group) who definitely switch lanes.

Figure 11.5 Hypothesized relationship between response rate and threshold level in high and low time-pressure situations

road under changing levels of time-pressure. A total of 147 students from Ben-Gurion University served as subjects in the experiment. They were divided into three groups, each of which was assigned a different reaction time condition, representing a different level of time-pressure. Subjects were told that they always drive on lane *A*, and 12 choice situations, representing a mix of av–av and ap–ap conditions, were projected in a random order. Subjects were allowed 5, 10 or 25 seconds, depending on the group, to mark their probability (0–10) of staying in the current lane or switching to the other lane (*B*). The av–av and ap–ap conditions represented decision rules with varying compensatory levels. These data were used to examine the hypothesized relationship discussed above.

Thresholds for Lane Switching

An empirical test of the hypotheses and the estimation of the consequent threshold levels are presented in Figure 11.6. As assumed, both hypotheses were found to be significant, at least at the 0.07 level and had a minimal correlation coefficient of 0.8. Results for the av–av situations were slightly better, as one might expect in situations with less time-pressure. The theoretical assumption that under higher time-pressure decisions will be more risky

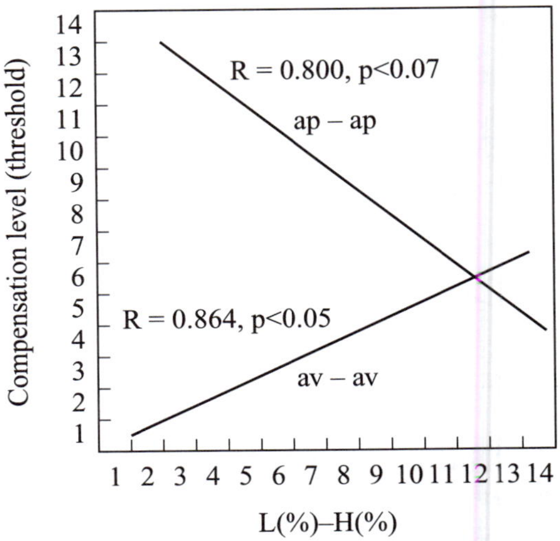

Figure 11.6 *Estimated levels of lane-switching threshold under high and low time-pressure*

and less accurate is indeed reflected by the lower significance and the smaller correlation found in ap–ap situations. The compensation level presenting the threshold is ordinal and therefore it is only a proxy; nevertheless, it shows the significant decrease of the threshold as reflected by the increase in the number of people who responded (changed lane). This finding suggests that the respondents' inhibitory criterion was smaller than the momentary value of their preference state. An increase of the threshold is respectively noticed under less, or no, time-pressure.

6. CONCLUSIONS

The various empirical results presented throughout the chapter correspond with its theoretical conceptions. The measurability of the behavioural threshold changes according to the situation. *Under low, or no, time-pressure*, the threshold can be calculated subjectively in measurable terms such as, for example, the expected delay time. Nearly always, the threshold expressed this way for activity behaviour is 30–60 minutes. The threshold is, however, likely to change only by trip purpose. Evidently it is smaller for non-work trips because people have fewer obligations in these situations.

Under high time-pressure, the threshold cannot be calculated in terms of time gain and is not defined in exact quantitative terms, but rather as a subjective construct affected by the exogenous variables of decision field theory. Evidently, however, more people are likely to have smaller thresholds under high time-pressure than under lower time-pressure. This suggests further research in which we will measure quantitatively the effect of the theoretical construct on the subjective choice probability. Such a measure is usually possible through the Heise Sheaf Coefficient (Heise, 1972) within a path-analytic scheme.

REFERENCES

Adler, J.L., W.W. Recker and M.G. McNally (1993), 'A Conflict Model and Interactive Simulator for Predicting Enroute Driver Behavior in Response to Real-Time Traffic Condition Information', Transportation, **20**, 83–106.

Busemeyer, J.R. and J.T. Townsend (1993), 'Decision Field Theory: A Dynamic–Cognitive Approach to Decision Making in an Uncertain Environment', *Psychological Review*, **100**(3), 432–59.

Doherty, T.S., and W.K. Axhausen (1998), 'A Conceptual Model of the Weekly Household Activity–Travel Scheduling Process', paper presented at the Network on European Communications and Transport Activities Research Euroconference, Israel.

Heise, D.R. (1972), 'Employing Nominal Variables, Induced Variables, and Block Variables in Path Analysis', *Sociological Methods and Research*, **1**, 147–73.

Hutchinson, R. and C. Dudeck (1979), 'Delay, Time Saved, and Travel Time Information for Freeway Traffic Management', *Transportation Research Record*, **722**, 36–9.

Khattak, A.J., L.J. Schofer and S.F. Koppelman (1993), 'Commuters' Enroute Diversion and Return Decisions: Analysis and Implications for Advance Traveler Information Systems', *Transportation Research A*, **27**(2), 101–11.

Koppelman, S.F., R.C. Bhat and L.J. Schofer (1993), 'Market Research Evaluation of Actions to Reduce Suburban Traffic Congestion: Commuter Travel Behavior and Response to Demand Reduction Actions', *Transportation Research A*, **27**(5), 383–93.

Mahmassani, H., C. Caplice and C. Walton (1990), 'Characteristics of Urban Commuter Behavior: Switching Propensity and Use of Information', *Transportation Research Record*, **1285**, 57–69.

Mahmassani, H. and J. Rong-Chang (1998), 'Bounded Rationality in Commuter Decision Dynamics: Incorporating Trip Chaining in Departure Time and Route Switching Decisions', in T. Garling, T. Laitila and K. Westin (eds), *Theoretical Foundations of Travel Choice Modeling*, Oxford: Elsevier Science Ltd.

Novaco, R.W. et al. (1990), 'Objective and Subjective Dimensions of Travel Impedance as Determinants of Commuting Stress', *American Journal of Community Psychology*, **18**(2).

Stern, E. (1996), 'Congestion Response Behaviour in Tel Aviv Metropolitan Area', in Y. Gradus and G. Liphshitz (eds), *The Mosaic of Israeli Geography*, Beer Sheva: Ben-Gurion University Press, pp. 147–58.

Stern, E. (1998), 'Travel Choice in Congestions: Modeling and Research Needs', in T. Garling, T. Laitila and K. Westin (eds), *Theoretical Foundations of Travel Choice Modeling*, Oxford: Elsevier Science Ltd, pp. 173–200.

Stern, E. (1999), 'Reactions to Congestion under Time Pressure', *Transportation Research C*, **7**(2–3), 75–90.

Stern, E. and Z. Nasrin (1994), 'Traffic Congestion and Residential Relocation', *Proceedings*, The International Geographical Union Regional Conference, Prague.

Stern, E., P.H.L. Bovy and M. Tacken, (1998), 'Traffic Congestion and Behavioural Reactions Policy Implications Exemplified for the Dutch Case', in K. Button, P. Nijkamp and H. Priemus (eds), *Transport Networks in Europe: Concepts, Analysis and Policies*, Cheltenham: Edward Elgar, pp. 103–22.

Tacken, M. and E. De Boer (1989), *Flexitime and the Spread of Traffic Peak Hour – An Analysis of Conditions and Behaviour*, OSPA, Delft: Delft University of Technology (in Dutch).

Tacken, M. and E. De Boer (1991), *Changes in the Spread of Travel and Working Times due to Opening of the Amsterdam Orbital Motorway*, OSPA, Delft: Delft University of Technology.

12. A conceptual model of the weekly household activity/travel scheduling process

Sean T. Doherty, Eric J. Miller, Kay W. Axhausen and Tommy Gärling

1. INTRODUCTION

Over the past several decades a strong argument has been made for the use of an activity-based approach to further our understanding of travel behaviour, to improve travel demand forecasting, and to better assess the impacts of emerging transportation policies. The rationale for an activity-based approach has been well documented (for example Ettema and Timmermans, 1997). One of the key questions of this approach is how individuals and households make and adapt their activity/travel decisions. These include the interdependent decisions about which activities to perform, where, at what time, for what duration, with whom, coupled with mode and route choice. When these decisions are coupled with their planning and execution over time, they define an 'activity-scheduling' process. This series of dependencies can be viewed as follows:

Scheduling process
⇑
Activity–travel pattern (or schedule)
⇑
Activities
⇑
Trips

As one moves up in this framework, a greater understanding of trips and travel patterns is achieved, especially of the more complex trip chaining, off-peak and discretionary trips. The trade-off in understanding comes at a price in terms of the complexity of the phenomenon and of the observation task.

Regardless of the added complexities, travel behaviour researchers are increasingly recognizing the need for in-depth research into the household activity-scheduling process. Early on, Pas (1985) noted that existing theories and methodologies dealt almost exclusively with travel and related behaviour at particular points in time, but that 'understanding travel and related behaviour requires the development of models of the process by which travel and related behaviour change'. Jones et al. (1990) noted that 'household responses to changes in transport and land use supply characteristics may be varied and complex. They include reorganization of trips into tours, reassignment of trips/tasks among household members and complex rescheduling of activities and travel. Conventional models of travel demand consider only simple adjustments in the daily travel pattern.' Axhausen and Gärling (1992) emphasize, in general, that the rescheduling of activities is at the core of many of the changes in travel behaviour brought on by recent policy initiatives related to information technology and transportation demand management. Thus it is becoming ever more important that the development of travel forecasting models capable of assessing these types of emerging policies needs explicitly to account for how people would temporally and spatially adjust their travel behaviour, which is dependent on an underlying process of activity scheduling.

Despite this realization, previous modelling efforts have relied mainly on the traditional utility-maximization framework to replicate specific aspects of the scheduling process in isolation or in limited combinations. Other models have attempted to capture the choice of an entire daily activity pattern. However, the behavioural validity of the utility-maximization framework as a description of how people actually make decisions has continuously been questioned (for example Gärling, 1998) and insights from cognitive psychology about how people perform complex scheduling tasks suggest that people apply a large range of heuristics and strategies when faced with such tasks (Payne et al., 1993). Gärling et al. (1994) argue that an even more serious issue relates to the tendency of these models to be confined to specifying what factors affect the final choice of *pattern* whereas the *process* resulting in the choice 'is largely left unspecified'.

In response to these criticisms, recent modelling efforts have attempted more explicitly to replicate the sequencing of decisions made during the scheduling process under alternative behavioural structures. These include those that involve production systems (for example Gärling et al., 1994), a mixture of utility-maximization and heuristic rules (for example Ettema et al., 1993), as well as those involving a micro-simulation approach (for example RDC Inc., 1995). Far more extensive reviews can be found in Axhausen and Gärling (1992), Ettema and Timmermans (1997), and Bowman and Ben-Akiva (1997). What is important in this context is that these approaches, to varying degrees, attempt to limit the assumptions made about the underlying scheduling behaviour of individuals and households so that models can be used to assess the impact of policy initiatives that previous models were simply incapable of doing. However, as pointed out in the review by Bowman and Ben-Akiva (1997), even these models 'can be challenged as to the validity of its decision protocol', noting that in each model reviewed, 'specific assumptions about how the decision-maker goes about the search and decisions are structured into the simulation. These assumptions may be wrong in enough cases to invalidate the model's parameter estimates and predictions.'

The challenge of activity scheduling is thus threefold: (a) the details of the process are largely unknown, particularly in terms of how schedules are formed within a household over time; (b) its complexity makes the observation task very challenging; and (c) even with full information, the resulting modelling tasks require the piecing together of at least the basic dimensions of the activity-scheduling decisions that account for how they are formed and potentially modified. What seems clear is that existing activity diary techniques can provide only a limited amount of information on the underlying process, and without further empirical work, activity schedule modelling will continue with strong unsupported behavioural assumptions that limit their potential.

2. DATA COLLECTION METHODOLOGY

The week-long Computerized Household Activity Scheduling Elicitor (CHASE) survey addresses the problems of collecting data on the underlying household activity-scheduling process. Considerable amounts of testing were devoted to the development of the CHASE survey, and a previous paper has been devoted to a description and assessment of the design based on a sample of households (Doherty and Miller, 2000). Only a brief summary of its design is given here in order to provide a basis for the empirical results presented.

CHASE goes beyond previous methodologies (Hayes-Roth and Hayes-Roth, 1979; Ettema et al., 1994) by providing a means to observe the scheduling process as it occurs in reality in a household setting over a multi-day period (as opposed to an individual in a laboratory setting over one day). In this way it is able both to capture routine and complex scheduling processes as well as to observe those scheduling decisions made during the actual execution of the schedule that come as a result of unexpected and impulsive events, continuing changes in planning decisions, and interaction with the environment and other people with whom activities take place jointly.

The CHASE programme is designed to track the sequence of steps whereby activities from a household 'agenda' are added, deleted and subsequently modified during their execution to form household weekly activity schedules. An upfront interview is used to establish a household's activity agenda, which consists of a full list of activities potentially performed by household members, along with their attributes. This information is entered by an interviewer into computerized 'forms' linked to a database file that the CHASE program can access in order to display the information back to the user in choice situations.

Adult household members are shown how to use the CHASE software program on a laptop computer, which is subsequently left in the household for a week-long period (users begin recording their schedule decisions on a Sunday evening for activities that cover the following Monday to Sunday). Parents complete their children's schedule. Users are basically instructed to log in daily to the program, and continuously add, modify and delete activities to an ongoing display of their weekly schedule. A household member's weekly schedule is displayed as a series of columns depicting the days of the week from Monday to Sunday, and a series of rows depicting 15-minute time blocks starting at midnight, not unlike a typical day planner. Scrollbars allow the user to adjust the view. For parents, tabs along the left side of the screen allow them quick access to their children's schedules. Each activity on the schedule appears as a series of 15-minute coloured blocks, with the first of these boxes normally displaying the activity type and location. Multiple short activities in the same time block will scroll horizontally.

The main menu for the CHASE program includes the options to add, modify or delete an activity. To add a new activity, the user first selects a series of time blocks using the mouse, then selects the add command. This brings up a dialog box that prompts the user for the activity type, location, mode, travel time and involved persons. For the most part, the user simply points and clicks his/her choices using pull-down lists and option boxes. To modify an existing activity, the user simply selects it on screen and chooses the modify menu command. This brings up the same dialog box within which the user can modify any of the previously entered attributes. Aside from these basic scheduling options, the program automatically prompts the user for all additional information. This includes prompts necessary to track the scheduling process accurately (for example when and why they made certain choices), prompts for other information of interest (for example costs of activities) and prompts designed to encourage accurate and timely completion of the schedule.

A total of 41 households (66 adults, 14 children) completed the survey during the months of April–June 1997. Households were recruited via advertisements on the McMaster University campus in Hamilton, Ontario, and were offered $50 to participate. Fourteen of these households were married couples, 13 were married couples with young children aged between 2 and 10 (six with two children, seven with one), and 14 were single-person households. Adults were either working full-time (35) or part-time (4), conducting post-graduate research full-time (19) or part-time in conjunction with other part-time work (4), or were undergraduate students (4; 3 of whom had part-time jobs). The majority of households were located within two kilometres of the McMaster University campus, which is situated at the very tip of the western end of Lake Ontario (Hamilton Region population: ~600 000). A sub-sample of 55 adults was used for the analysis presented in this chapter.

The results presented in Doherty and Miller (2000) showed that the pre-definition of a household agenda, coupled with a familiar and graphical computer-based survey design, proved to be a practical approach to gathering a wealth of information on the underlying process, while minimizing the burden on respondents. In summary, it was found that adults spent an average of 16 minutes per day to complete their own schedules, and 9 minutes per day for their children. The log-in durations were found to remain at a consistent length after the initial period of scheduling. Only about 10 per cent of the survey days were skipped, mostly on or near weekends when respondents were out of town. During the week, adults made an average of 12.1 add, 2.4 modify and 0.6 deletion steps per day to schedule 12.4 activities and 4.9 trips per adult per day. Taken together, the total number of activities and trips reported per adult per day from Monday to Thursday was very stable at 17.9, 18.0, 17.9, and 18.1 respectively. Friday and Saturday were comparable at 16.8 and 16.9, followed by 15.7 on Sunday. Thus, in addition to a low respondent burden, the survey appears to have minimized the potential of multi-day surveys to provide less information on subsequent days.

This approach goes a long way towards solving the data collection problem highlighted by Bowman and Ben-Akiva (1997) that simulation models of activity scheduling (citing Ettema et al., 1993) require 'very complex surveys for model estimation' wherein 'respondents must step through the entire schedule building process'.

3. CONCEPTUAL MODEL DEVELOPMENT

The weekly household activity scheduling process model presented below is supported by empirical analysis and by various flow diagrams that provide a visual means to connect the various components of the model (see Figures 12.1–12.3). Although the ultimate goal is to implement the model as a computer algorithm, this is not necessarily a binding restriction on the design. Note that *italicized* terms in the text refer directly to components of the model in the figures.

First, it should be stressed that activity scheduling is a behaviourally complex process that spans many time horizons and individuals. This has not only

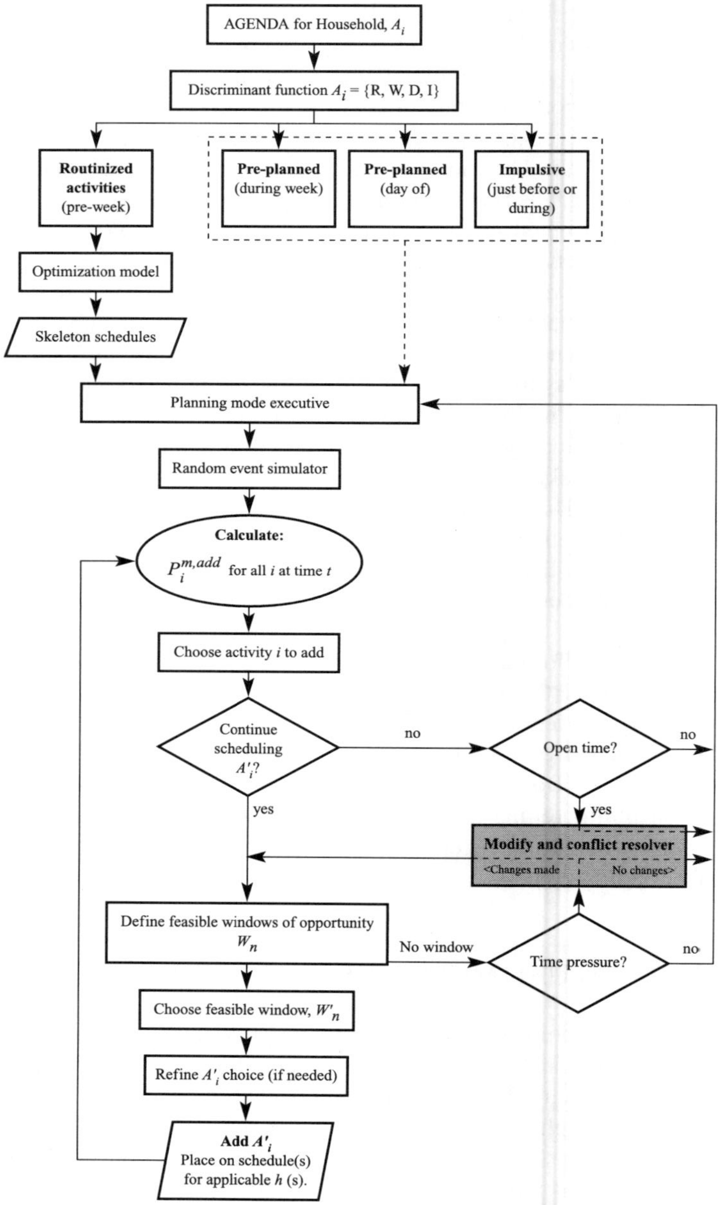

Figure 12.1 Main model for weekly household activity scheduling process

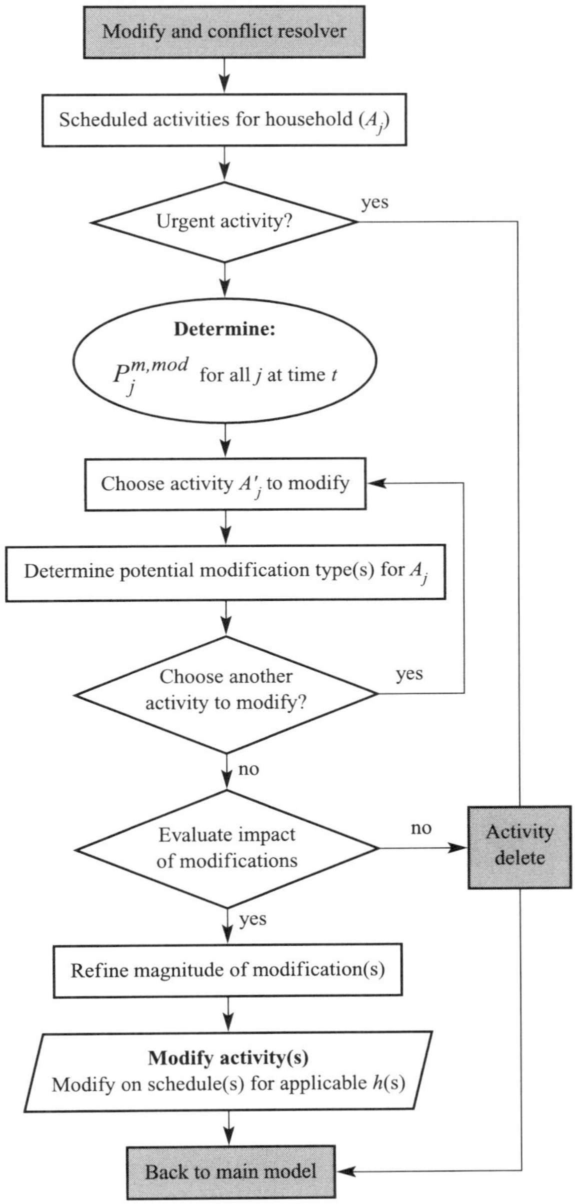

Figure 12.2 Modify and conflict resolver sub-model

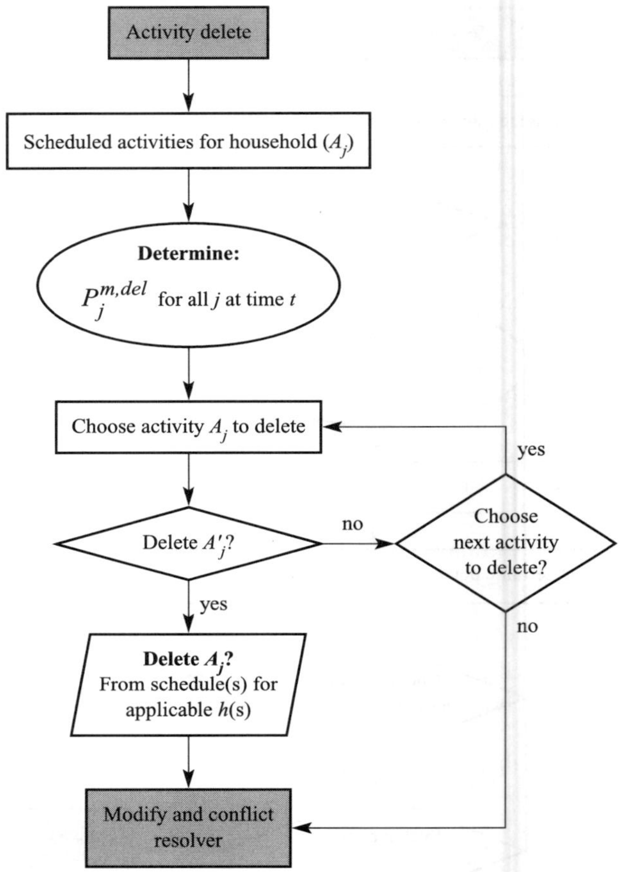

Figure 12.3 Delete sub-model

required innovative observational tools, but also complex modelling structures. However, the trade-off comes in an improved understanding of and ability to forecast complex travel behaviour that would not be possible without a knowledge of the underlying scheduling process that gives rise to this complexity.

Basic Structure

To assist with an understanding of the overall model design, a summary of its basic structure is provided here. In general, the model attempts dynamically to replicate the scheduling process as it occurs over time through the use of various modelling constructs and decision rules run in sequence. It begins by taking a household's *weekly agenda* of activities, and establishes a set of *routine activities* and a *skeleton schedule* for the week. This is followed by scheduling decisions (additions, modifications, deletions) made during execution of the schedule. These include pre-planned decisions, decisions made on the day, impulsive decisions, and decisions that result from random events. A *planning mode executive* controls the flow of decisions, which inherently involves the movement through time. Decisions about what activities to schedule at any given moment are determined by the priority of activities on the agenda. Priority is determined as a function of static activity attributes from the agenda and dynamic aspects of the schedule at the particular moment. Once an activity is chosen for scheduling, a feasible window of time is chosen, and refinements in the activity made before it is placed on the schedule for a particular household member(s). Refinements include any needed duration, timing, location, or mode choices. In many cases, these refinements will be highly constrained given the relative fixativity of the activity in time and space and/or the constrained nature of the circumstances, making these choices rather straightforward.

Conflicts that arise due to random events or time pressures, or cases where activities may be extended to fill time, are handled by the *modify and conflict resolver*. This procedure takes the previously scheduled activities and determines those most likely to be modified. A set of possible modifications is determined, and a choice is made as to which ones to implement and to what extent. If the (set of) modification(s) does not meet the requirements of solving a conflict, then the *deletion of an activity* is considered. The procedure for deletion is similar to modification, except that the activity attempting to be scheduled is compared directly to the revised priority of the activity chosen for potential deletion. If none of the deletions is justifiable, then the model reverts to the beginning, and the originating activity is left unscheduled. If an activity is deleted, control reverts to scheduling to assess the new window feasibility.

The following sections describe the model components in more detail.

The Household Activity Agenda

On a fundamental level, activity scheduling reflects personal and household-related basic human needs constrained in time, capability and in space by the urban environment. These needs can be viewed as manifested in a household's activity *agenda* which represents the initial input to the main model, as shown in Figure 12.1. A simplified example weekly agenda is provided in Table 12.1. The agenda consists of a list of uniquely defined activities that a household could potentially perform, including highly infrequent activities

Table 12.1 A simplified example of a household weekly agenda

Activity	Applicable household members	General location	Attributes			
			Duration (hrs) (mean)	Frequency (per week)	No. of perceived locations	Etc.
Work	Male head	Home	2	2	1	
Work	Male head	Out of home	8	5	1	
School	Child	Out of home	8	5	1	
Grocery shop	Female head	Out of home	1	3	12	
Grocery shop	Male and female	Out of home	2	1	12	
Active sport	Male head	Out of home	1	1	2	
Activity sport	Male or female	Out of home	2	1	1	
Chauffeuring	Male or female	Out of home	0.5	5	1	
Socializing	Male or female	In or out of home	3	2	10	
Etc.						

that are only occasionally performed to fill up free time. Each activity on the agenda is viewed as having a unique set of (perceived) attributes that affect their scheduling, including duration (min, max, mean), frequency, time limits, involved persons, costs, perceived locations, and so on. If an activity constitutes several unique subtypes that differ substantially in their attributes, then they should be defined separately (for example work at the office, work at home; minor grocery shopping, major grocery shopping, clothing shopping). What is key to the success of the scheduling model is not the activity types as defined by traditional means (for example work, school, shopping, mandatory, discretionary, and so on), but rather their salient attributes that serve to explain how they are scheduled in the context of an ongoing process. This gives the model the ability to address any number of individuals/household types, regardless of their employment status or preferences.

Although the derivation of household activity agendas is of considerable interest on its own, it is taken as exogenous to the process of scheduling in the short term. It is assumed that household agendas in a given urban area can be separately simulated or otherwise derived at regular time intervals as required by the scheduling model. The process by which these agendas are derived is an important component of the overall model, but is beyond the scope of this chapter. Our suggestion at the moment is that the agendas need to be based on key socio-demographic characteristics of the household, household resources and constraints, and environmental factors. Attributes such as duration and frequency could be based on existing activity diary data, whereas the set of perceived locations would need to be simulated with a cognitive model using inputs such as the location and length of stay at current residences and employment locations. Incorporating the learning of new locations, new activity types, and the modification of activity attributes over time would be an important aspect of this model. An argument for linkages between the agenda and scheduling via learning processes in the long term is an important future consideration (for example as you execute your schedule you become aware of new locations for activities that should be added to the agenda).

Routine Weekly Activity Skeleton

Empirical evidence derived from the CHASE survey shows that households begin the week with a firmly established set of *routine weekly activities*. From Figure 12.1 (looking only at the 'additions' made on the 'first Sunday') it can be seen that about 45 per cent of the activities that take place on weekdays, and about 20 per cent of those that take place at the weekend are pre-planned before the week starts on or before the first Sunday (totalling approximately 34 activity additions on the first Sunday). The remainder of activities are planned as the weekly schedule is executed (discussed in the next section). Of the decisions made on the first Sunday, a full 70 per cent were part of multi-day entries (the activity was added on 2 or more days simultaneously), with 80 per cent of these consisting of entries across 4+ days. Comparatively, on Monday, only 21 per cent additions were part of multi-day entries, followed by 2 per cent, 6 per cent and no more than 1 per cent on remaining days of the week. Such entries are indicative of highly frequent and routine activities, as suggested.

Further evidence suggests that other key factors differentiate the activities pre-planned on the first Sunday besides their routine nature. First, they tend to be of a longer duration then subsequent days, averaging 208 minutes long for adults in the sample. This was significantly higher than activity durations that were pre-planned one or more days in advance during the week (110 minutes), planned the day of (79 minutes) and planned impulsively (72 minutes), as confirmed by two tailed t-tests ($p<<0.0001$ in all three cases).

A rough examination of the activity types scheduled on the first Sundays also suggests that more highly fixed activities in space and/or time tend to be pre-planned before the weeks starts (for example work, chauffeuring, sleeping, sports events). Further empirical analysis using an appropriate discriminant analysis technique will be performed to further differentiate these types of

Table 12.2 Potential explanatory variables in the momentaneous priority model ($P_i^{m,s}$)

Characteristics/ Attributes	Indicator variable*
Attributes of Activities on agenda X_i	
Time window (feasible)	Earliest start – latest end time
	No. of Feasible days
Temporal flexibility	(Minimum duration)? (Time window size)
	(Frequency per week)? (No. of days occur on per week)
	(Days it occurs per week)? (No. of days it could occur on)
Duration	Minimum, maximum, mean
Frequency	Days between performance (observed or stated)
Activity direct costs	{Yes, No}
Joint scheduling	{joint mandatory, joint optional, individual}
Activity type by activity motivation/need	Categories: {physiological, institutional, social, psychological, household obligation, household task}

Table 12.2 concluded

Characteristics/ Attributes	Indicator variable*
Scheduling State at time *t*, S_i^t	
Schedule flexibility/window	(Minimum A_i duration)? (Maximum feasible window available)
Alternative household member flexibility	As above, for alternative household member
Activity history	(Days since last performance of A_i)? (A_i frequency)
	(Duration of last performance) ? (Mean duration)
	Number of activities (of same basic type) performed recently
Activity future	{A_i not scheduled, scheduled}
	(Days to scheduled performance of A_i)? (A_i frequency)
	Number of activities (of same basic type) scheduled for future
Activity habit history	Frequency of performance directly before/after given scheduled activity
Travel time	Total travel time for day
Spatial Attributes, M_i^l	
Spatial fixity	No. of perceived locations
	{only at home, only one out of home; in or out of home}
Perceived distance	Distance to nearest perceived location for A_i
Travel time	Perceived travel time to nearest activity location by fastest available mode
Scheduled activity attributes	Perceived distance in space between nearest activity before/after desired activity
Environmental Attributes, *E*	
Opening hours	Opening hour restrictions for A_i
Location attributes	Attraction variables for locations for participating in A_i

Note: * Values in { } indicate categorical variables.

activities from other activities on the agenda based on their key attributes, including duration, frequency, and indicator variables of temporal and spatial fixativity (see also Table 12.2 for further details on these variables).

It is plausible to assume that these routinized activities pre-planned before the week starts are the result of a long-term deliberate planning and adaptation process controlled by negative feedback. Thus they may represent an optimized activity pattern around which other scheduling decisions are made during the week. Given this, it is reasonable to assume that an *optimization model* would be appropriate to derive the *weekly skeleton schedules*. This model would use the discriminated activities as input, which represent a much more limited choice set of activities that more closely match the assumptions on which these models are based (that is, that activities are scheduled in an optimal way). Potential modelling techniques include those that start by generating all possible feasible combinations of skeleton structures and choosing the most optimal of the set (for example Recker, 1995), or the more recently developed neural network models that offer a more behaviourally realistic modelling technique (for example Kitamura and Fujii, 1998). The details of this model form, and how it would deal with a weekly optimization problem, are however left as a long-term research goal.

Weekly Scheduling Process

The remainder of this chapter focuses on the more deliberate scheduling decisions made during the week as the schedule is executed. After the first Sunday, adults consistently made about eight additions, two modifications, and one deletion per day during the execution of their schedule over the course of the week. These steps were taken to complete their schedules, which include an average of 12.4 activities and 4.9 trips per adult per day.

Figure 12.4 shows that the scheduling decisions applicable to each day are made on various time horizons. Outside of the routine activity additions made on the first Sunday (38 per cent overall), a substantial proportion of additions are scheduled impulsively just before execution (28 per cent overall), on the same day (20 per cent overall), or are planned one or more days in advance (15 per cent overall). The amount of pre-planning differs by day of the week. Most notably, more impulsive additions decisions occur on Saturdays, whereas more 'day of' decisions are made on Sundays.

Modification decisions exhibit a significantly different pattern from additions. The vast majority tend to be made impulsively (62 per cent overall), compared to on the same day (24 per cent overall) and planned more than a day in advance (14 per cent overall). This pattern is fairly consistent across the days of the week, as shown in Figure 12.4. Compared to modifications, more advanced thought appears to be put into deletions as a higher propor-

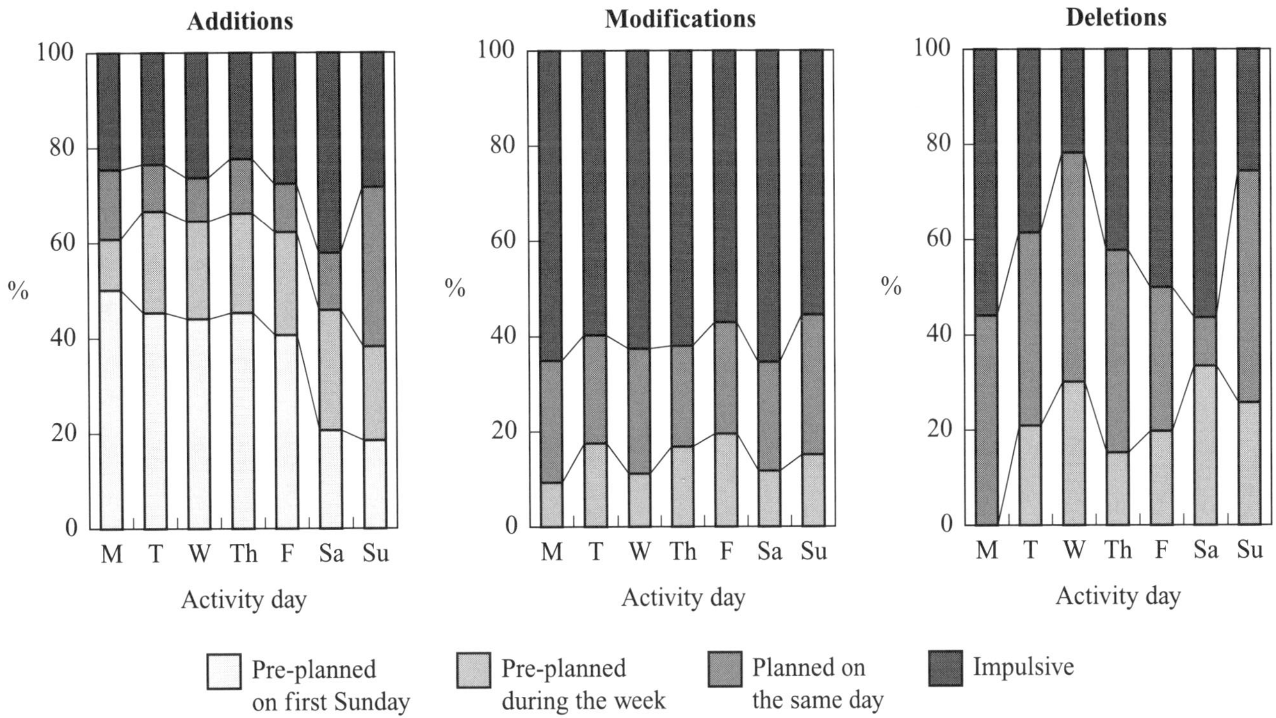

Figure 12.4 Proportion of additions, modifications and deletions made for each day by when they were planned

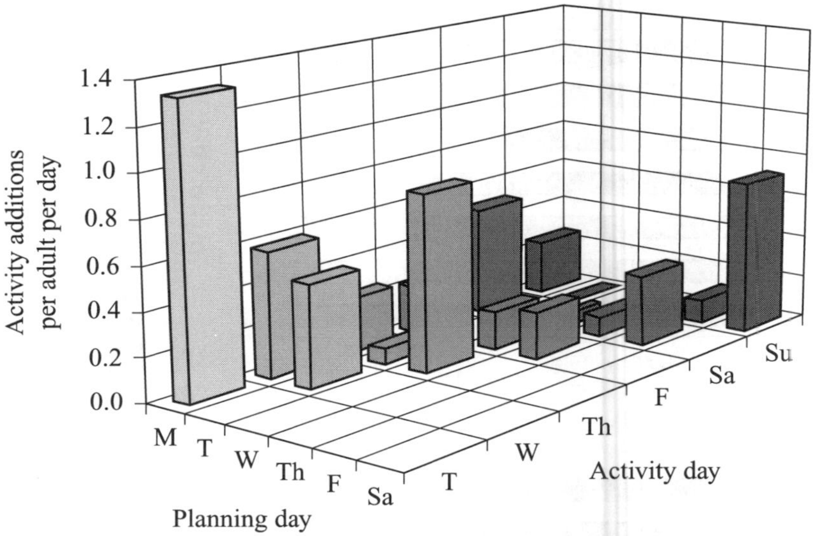

Note: * Restricted to activities planned more than a day before the activity occurred, exclusive of the first Sunday.

Figure 12.5 Pre-planned activities by planning day and activity day*

tion are made on the same day (38 per cent overall) and a lower proportion made impulsively (41 per cent overall).

Figure 12.5 shows in three dimensions the distribution of activity additions planned on one day, by the activity day they are planning for (exclusive of the first Sunday, which is shown in Figure 12.4). It clearly shows that in addition to pre-planning for the next day (62 per cent overall) people also reach out and make additions on future days as the week progresses (38 per cent overall), in an opportunistic fashion.

This evidence strongly suggests that activity scheduling is a dynamic process reflecting continued additions and revisions to a schedule over time. Scheduling represents a mix of routine followed by continued pre-planning and impulsive decisions made over the course of the week. This differs substantially from the notion that activities are planned and carried out in sequence all at once. Thus, if the goal is to develop a behaviourally sound model, then a dynamic model is needed – one that can simulate the fundamentally different types of decisions that occur over time.

The conceptual model presented in Figure 12.1 to Figure 12.3 attempts to account for the complexities of the scheduling process with a mix of empirically derived functions and a series of decision rules that serve to sequentially simulate the construction of activity schedules over time. Activity-scheduling decisions including additions, modifications and deletions to each household member's schedule, are simultaneously implemented as a series of pre-planned and impulsive decisions. At the heart of the conceptual model is a 'momentaneous scheduling priority' function that drives that selection and inherent sequencing of activity choices at each stage in the process.

Momentaneous Priority

Activity 'priority' has been suggested in the past as an important dimension in the construction of sequential scheduling models, particularly as a determinant for the sequencing and choice of activities. Past researchers have suggested that activity 'priority' has a dynamic quality, wherein activities of different types have priorities that change over time in response to different situations (Axhausen and Gärling, 1992; Gärling et al., 1989, 1998). It has also been suggested that activity priority in scheduling is related to the relative flexibility and fixity of activities, wherein inflexible and highly fixed activities are chosen first for scheduling, followed by more flexibly and less fixed activities (Cullen and Godson, 1975; Kitamura, 1983). The current model proposes that much of the complexity of the scheduling decision process is related to the changing level of priority associated with activities that household members have on their agenda at any one moment in space and time. Estimating a function describing how priority varies with different factors over time is therefore at the heart of the conceptual model.

Although activity 'priority' has been proposed as the determining factor in the choice of activities to schedule in previous models (for example Gärling et al., 1989, 1998), it has remained a difficult attribute to operationalize because of its highly subjective and dynamic nature. Asking people to assess the priority of a list of activities is difficult not only because of a definition problem (how the researcher defines/explains what constitutes high versus low priority will largely affect the results), but because the priority of an activity depends on the situation. Any static assessment of the priority of activities will be inadequate to deal with all possible situations that arise during the scheduling process. This stresses the importance of differentiating between an activity's 'general/overall' level of priority (all other things being equal) versus its 'scheduling' priority (which depends on the situation). For instance, a person may tell you that being at home with their children is generally a higher priority than being at the pub with friends. However, under the right circumstances (for example having just spent the whole day with one's children, who are now sleeping, and the person hasn't been to the pub in a while), this person might be found at the pub. Any measure of priority must be able to account for these observed differences.

If 'scheduling priority' determines the choice of activities in a given situation, then any measure of it requires observations of activity-scheduling choices under a range of situations – precisely what the CHASE survey is designed to accomplish. The following modelling function of momentaneous scheduling 'priority' is thus proposed for development:

$$P_i^{m,s} = f(X_i, S_i^t, M_i^l, H, E) \tag{12.1}$$

where:

P_i: 'Priority' of activity i at scheduling mode m for scheduling operation s
m: Scheduling mode – pre-plan, day-of, and impulsive planning
s: Scheduling operation – add, modify, delete
X_i: Attributes of activity i on household agenda
S_i^t: Scheduling state characteristics of activity i at time t
M_i^l: Spatial possibilities for activity i at location l
H: Household characteristics
E: Environmental characteristics.

At any moment in the scheduling process over the course of a week, the priority of activities in the household agenda can be evaluated in terms of their relative priority, given values for the attributes in the model. Separate models would be constructed for the priority of activities for addition, modification or deletion to the schedule (represented by the subscript s). The form of the model is also proposed to depend on the mode (m) of scheduling (pre-planning, day-of planning, impulsive decisions). The exact form of the priority model and its estimation would require considerable exploration.

The explanatory variables in the priority model ($P_i^{m,s}$) include a range of static variables (A_i, H, E), and dynamic scheduling state variables (S_i^t, M_i^l) that continuously change over time (t) and space (l). The scheduling state variables are what makes this model unique so far, and give it the power to explain the apparent behavioural complexities of observed activity-travel patterns. The range of potential explanatory variables is outlined in Table 12.2. Of particular note are the history and future dependent variables that account for the likelihood that activities that have taken place recently or been planned for a future time would be associated with lower priorities. Also, the temporal and spatial fixity of activities is captured by a combination of activity and scheduling attributes. The attributes of other household members' schedules, such as the relative flexibility of their schedule at a particular moment, is also proposed as a determining factor in activity priority. Many other variables are proposed in the table to account for the role of habits, travel times and joint activities.

The behavioural power of the priority ($P_i^{m,s}$) model is threefold. First, because the model is sequential in nature, the scheduling state variables can be reassessed at each time step in the model after each decision. Certain activities will have a tendency to jump up in priority depending on the circumstances. This allows infrequent, discretionary, or otherwise unusual activities to emerge depending on the situation, contributing to complex activity/travel patterns. For instance, although a person may have a high-priority job to get to at 9:00 a.m. and possibly some high-priority shopping that needs to be done before the end of the day, s/he may be sitting in a coffee shop located on the way to work at 8:30 a.m. This may be because the person was faced with a short window of opportunity for the activity after having already dropped off the children at school at 8:15 and a close location, which allowed for an apparently flexible activity to emerge (coffee) as high priority. Shopping was assigned a low priority at that moment because of the short time window and because few perceived shopping locations were available in the vicinity.

Second, it simplifies the activity scheduling decisions – empirical evidence suggests that those decisions made before the week commences (on the first Sunday) tend to be highly routine, making the decision about their exact timing and location a relatively straightforward task. Those activities that are more flexible tend to be scheduled within an already constrained spatio-temporal environment limiting the specific choices of location, start–end times, duration, household members involved and mode choices.

Third, the momentaneous priority model ($P_i^{m,s}$) indirectly contributes to the sequencing of activities in terms of the order in which decisions are made (pre-planned vs impulsive), and the order of execution. This sequencing is not an explicit aspect of the model, but rather reflects the fact that impulsive decisions and, to a lesser extent, day-of and pre-planned decisions are made in light of the open time slots in the skeleton schedule established on the first Sunday. This implies a certain order among existing activities on a weekly, daily or time-window scale.

Sequential decision structure

Once the mode of scheduling is set by the *planning mode executive* (described in more detail in the section to follow), the model uses the same basic set of sequential decisions to handle random events, add high-priority activities to the schedule depending on the scheduling mode, and invoke modifications and deletions where necessary.

Ignoring for the moment random events, the choice of activity to add (*choose A_i to add*) is based on the momentaneous scheduling priority of all applicable activities on the agenda ($P_i^{m,s}$) for that particular mode of scheduling (s). Some logical rules may also be used to limit the choice set in certain circumstances due to constraints – or to increase the priority of certain activities directly. The exact method would depend upon the extent to which the priority model ($P_i^{m,s}$) accounts for the constraints via the inclusion of appropriate variables. For instance, the maximum size of any feasible time window (max W_n) on a schedule relative to the minimum duration of an activity (min d_i) should presumably influence its priority for scheduling. An appropriate variable for investigation might be:

$$\frac{\min d_i}{\max W_n}$$

The smaller this value is, the higher the priority should be. An example of a rule for excluding certain activities is:

$$\text{IF}\left[\frac{\min d_i + \min t_i}{\max W_n}\right] > [\beta]\ \text{THEN}\left[P_i^{m,add} = 0\right]$$

This rule operates by assigning a zero priority to activities that are simply too long to fit in a given time window, considering also any necessary travel time (t_i). The threshold value of β should be larger than 1 and could be determined empirically, or simply set to a pre-specified value.

Still further rules could be incorporated before the final choice to artificially increase the priority of certain activities (A_i) when dependent/related activities (A_j) have already been scheduled. For example, an out-of-home recreation activity for parents may first depend on a chauffeuring activity for the children (to a babysitter), or a socializing activity (going to the pub) may habitually follow the playing of a sporting event. These would, of course, depend on a well-defined agenda that establishes these dependencies.

Once the momentaneous priority is determined and constraints met, the choice of activity to schedule is made based on an appropriate decision rule (*choose* A_i *to add*). The simplest rule would be to select the activity with the highest priority. Alternatively, the choice could possibly be made randomly, for instance if there is time pressure. Selection of activities to schedule in this fashion proceeds until a threshold level of priority for the given mode m is reached. The decision rule used at this stage could be of the form:

$$\text{IF } P_i^{m,add} > (\alpha_m)\ \text{THEN } [\textit{continue}] \tag{12.2}$$

where the α_m threshold value is determined empirically for the given mode of scheduling (α_m would be relatively higher during pre-planning compared to impulsive planning, reflecting the notion that only high-priority activities are pre-planned). This rule implies that only the priority of the current activity in question is considered in the decision. An alternative would be to base the decision on the sum of priorities of activities on the agenda, replacing the left side of the above equation with

$$\sum^{i} P_i^{m,add}$$

This would reflect the aggregate amount of pressure the particular person is under to continue scheduling activities. If many of the activities that are typically pre-planned are not yet placed on the schedule, then the sum of activity would be high, invoking further scheduling until the threshold level is reached.

More precise decisions about scheduling are not made until an activity is chosen (A_i') and a decision to continue is reached. First, *feasible windows of opportunity* (W_n) must be identified. These are defined as open time periods of length at least equal to the minimum duration of the given activity that fall within the feasible hours for the given activity. Feasible hours are defined by the opening hours of any required facilities (if applicable) and any preferences of the individual reflected in the agenda (for example a preference for shopping only on weekends).

In the event that more than one window is feasible, another rule is needed to *choose the feasible window*. Given that it has already been established that the activity is of high priority, it may suffice to use a satisficing rule that the first available window in time be chosen. However, in some cases, other attributes of the windows may be considered, such as their length, or the attributes of activities that bound the window. It may be appropriate to develop a feasible window choice model, using attributes of the windows and the activity as determining factors.

Having chosen a high-priority activity (A_i') and identified a window of opportunity (W'_n) greatly constrains the remaining choices to a specific spatial–temporal setting. These include the planned choices of activity duration, location, start and end times, mode of travel and involved persons (*refine A_i' choice*). In many cases, these choices are already constrained enough that only one is feasible. This would be the case for many pre-planned activities that are highly fixed in time and space and involved persons, and for activities already impulsively planned within highly constrained situations.

For those activities where a choice on one or more dimensions exists, more definitive rules would be needed to make the refinement choices. Rules would be needed to identify what the alternatives are for each dimension, and second, how the choice is made among alternatives. The scheduling model would provide a means to do the former – that is, providing a means to limit the range of feasible alternatives to a manageable level, given the spatial–temporal setting of the decision and attributes of other activities in the vicinity. In the case of activity duration, it would naturally be constrained by the length of the time window available less any necessary travel time, and the maximum and minimum duration for the activity. An aggregate measure of the number and magnitude of other high-priority activities that may need to be scheduled as of yet may also be used to constrain the upper bound on duration. The location will already be constrained by the perceived location choice set in the agenda and travel time relative to where they are in space and where they may have planned to be at the end of the activity. Constraints on location related to the household (for example at least one parent at home with children in the evening) could also be formed. The modes available would be limited to what is currently available and the implications for travel time. The start time would probably be given from the end time of the preceding activity, whereas the end time would be governed by the duration and opening hours. Who would be involved, if a choice exists, would depend on the scheduling state of various household members. Operationalizing these constraints would require a series of 'if … then' rules to establish the choice set of alternatives. The actual rules then used to make the choice amongst the alternatives would need to be arrived at through more in-depth cognitive analysis of decision making, which is beyond the scope of the current chapter.

Once refinements in the activity are made, the activity is added to the schedule for the applicable household members (*ADD* A_i'). Control is then returned to the momentaneous priority model, which re-evaluates all activities on the agenda after each addition. Subsequent additions (and possible modifications/deletions) are then continued until a decision is made to halt scheduling in any particular mode (via *continue scheduling?*).

At specified times during the scheduling process, modifications and deletions may occur in response to *random events*, scheduling *time pressure*, or convenience (*open time*). In all cases, this leads to the *modify and conflict resolver* sub-model, although upon return, it may proceed in different directions, as shown in Figure 12.1. Each of these processes is described in the following sections.

4. RANDOM EVENT SIMULATOR (RES)

In the impulsive planning mode, the RES serves to randomly generate 'unexpected' events that require immediate attention for scheduling. Two main types of events may be generated. The first are unexpected changes in the duration of planned activities, generated each time the model moves to the beginning of an activity (A_j) in the impulsive scheduling mode (as determined by the *planning mode executive* – see section below for more detail). These events represent the distribution/flexibility of the activity's duration, as well as changes in travel as a result of unexpected congestion, route changes or mode changes. The extent of any duration changes would be constrained by some rules related to the fixativity of activities that follow the given activity – for instance, if a highly fixed activity follows a given activity, then an upper bound on any unexpected changes in duration may be set (for example if a recreational event precedes a fixed commitment to chauffeur one's children, then random duration changes should have an upper bound).

In cases where the unexpected change leads to a decrease in duration or an increase that does not cause any conflicts with other planned activities, the changes are made to the schedule. In cases where a conflict arises, the *modify and conflict resolver* is evoked. The behavioural response to such a situation is similar to that when faced with time pressure – one must now squeeze a high-priority activity into a time slot that is no longer adequate. To operationalize this behaviour in the model, the activity can be temporally removed from the schedule, while at the same time assigning a high priority to the activity along with fixed start and end times and duration. In this way, scheduling would proceed with the choice of this activity, to the identification of the window, to time pressure, and to the *modify and conflict resolver*, where the conflict would eventually be resolved due to its high priority. After any unexpected changes in duration are implemented, the activity is executed with no further changes.

The second type of random event concerns the simulation of urgent, emergency or surprise activities that are mostly outside the control of the person. These would be handled by directly assigning a very high priority to the activity item and having scheduling proceed. For immediate urgent activities (for example seek medical care), scheduling proceeds as is, in impulsive scheduling mode. For surprise activities that will occur at a later time (for example news of surprise visitors coming to your home the next day), the *planning mode executive* would be called upon to switch immediately to pre-planning mode to make the scheduling changes, then switches back to impulsive planning.

Time Pressure

Compared to random or unexpected events, *time pressure* is the result of a more purposeful choice process wherein the scheduling of an activity meets with insufficient time windows to conduct it. A decision rule is needed at this point to decide if the given activity priority is still high enough to warrant potential modification or even deletion of previously scheduled activities to accommodate its scheduling. Such a rule may be of the form:

$$\text{IF } [P_i^{m,add} >> \alpha_m] \text{ THEN } [\textit{Modify conflict resolver}]$$

Other more complex rules may be necessary that replace the right-hand side of this rule with an aggregate measure of the priority of existing activities (A_j) in the vicinity of the new activity (A_i') that may need to be modified. These latter priorities would be re-evaluated for the given moment, as opposed to using their priority level when originally scheduled. The end result is that either the *modify and conflict resolver* searches for a means to increase the size of the window, or scheduling returns to the *planning mode executive* and no further additions are attempted.

Open Time

Modifications may also be due to convenience, in a sense that a person has *open time* slots and the freedom to extend the duration of previously planned activities or choose a different location/mode in light of the opportunity. This type of modification would occur during impulsive scheduling in the event that none of the activities on the agenda merits a high enough priority for addition to the free time slot (that is, *continue scheduling* A_i'*?* No). The *modify and conflict resolver* is called upon to fill up the *open time*. This serves to maintain the continuity of time in the schedule (that is, a person must always be doing something).

Planning Mode Executive

The result of the momentaneous scheduling priority model is used to choose activities for addition, modification and deletion to household members' schedules in a sequential fashion. The sequential movement through time is simulated by the *planning mode executive* by controlling movement from one planning mode to the next. During the course of the week, people alternate between pre-planning activities for future days, pre-planning activities for the same day, and impulsive decision making. Weekly pre-planning mode is defined as those scheduling decisions made one or more days in advance of the event. Daily planning mode involves decisions made the same day as the event occurs. Impulsive decisions are made just before or during the activity.

The exact sequencing of pre-planning and impulsive decisions is unknown and remains a difficult phenomenon to disentangle, even with more repeated observation periods. For modelling purposes, one simplifying assumption would be to make all pre-planning decisions before the start of each day, wherein activities are planned for the same day or for a future day depending on the windows of opportunity. This would then be followed by impulsive decisions made at the start and end of each scheduled activity, starting with the first activity for the day. However, it may be argued that pre-planning is revisited during the day, perhaps as needed depending on the state of the schedule.

To review briefly, the model begins with a skeleton schedule for the week, followed by pre-planning during execution of the schedule, followed by impulsive decisions that occur in between each activity. Thus, once pre-planning halts for the day, the model proceeds in time via the *planning mode executive* to the starting point of the first activity, invoking the impulsive mode of planning. This mode is subsequently invoked at the beginning of each scheduled activity until a pre-planning mode is invoked by the Planning Mode Executive. The model inherently moves through time after impulsive decisions conclude at the beginning of an activity, and the model moves to the end of the activity.

Specifically, in the impulsive mode, the model proceeds to the beginning of the first scheduled activity for the day and first determines the actual duration of the activity via the *random event simulator*. This may result in changing the actual duration from the planned activity. Following this, the model proceeds in time to the end of the activity. At this time point, an empty time slot may or may not exist, and a decision is made impulsively to add an activity before the start of the next activity. The choice of activities to add during impulsive planning follows the same procedure as in pre-planning (including the possibility of the generation of a random new activity), except in the handling of empty time windows – whereas empty time windows are permissible during pre-planning, any *open time* windows must be filled during impulsive scheduling. In the case where none of the activities on the agenda meets the threshold for scheduling set for impulsive planning (*continue scheduling?* No) then a situation arises in which *open time* exists. The *modify and conflict resolver* is then invoked to revisit existing activities (A_j) to choose for modifying in order to fill up the time (see below for details), potentially resulting in a longer duration or new location/mode choices. In either event, the open time slots must be filled before the scheduler moves in time again to the start of the next activity (which may be an activity that was just added), and the procedure repeats itself. In this way, the final executed schedule is formed.

A graphical example of the above process is provided in Figure 12.6. The example shows how a person's schedule for a particular period of time would be sequentially pieced together by the model, from weekly skeleton to final executed schedule. Each time step is controlled by the *planning mode executive*. First, the skeleton schedule for the given time period starts with just two activities – work (A_1), and a routine household obligation (A_2). Next, a meeting is pre-planned a day in advance (A_3). This pre-planning is followed by a procession of impulsive decisions and random/unexpected events that occur

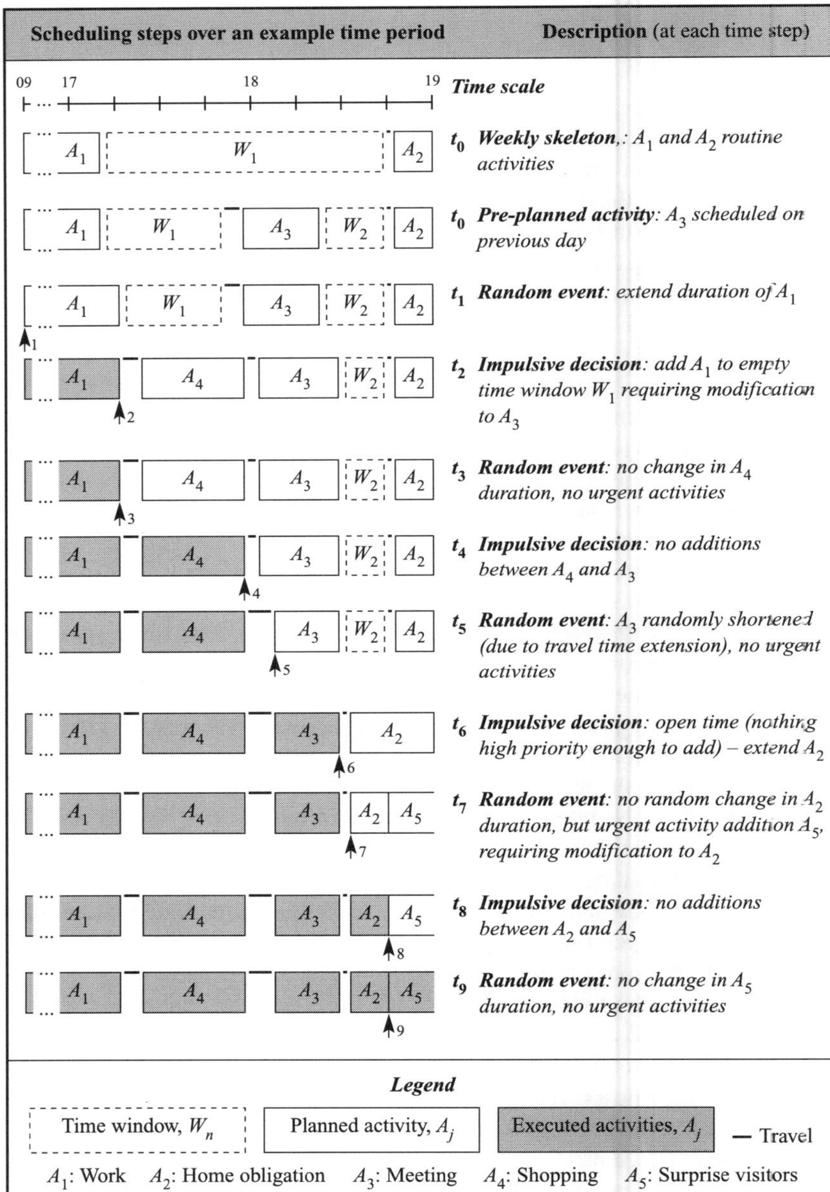

Figure 12.6 Graphical example of the sequential steps simulated by the model

in between each activity in time times steps of t_n. Each of these decisions serves to define the end state of the schedule and resulting activity-travel pattern.

Modifications

As described in previous sections, random events, time pressure, and open time could potentially lead to the *modify and conflict resolver (MCR)*, a flow diagram of which was presented in Figure 12.2. Except in the case of open time, the inputs are the list of scheduled activities (A_j), the new activity (A_i'), and the attributes of the chosen time window (W_n') in which the new activity does not currently fit. The minimum time necessary can be taken as the difference between the minimum duration for the activity (including travel time to the closest possible location) and length of the chosen time window. The option to bypass the MCR and proceed directly to the *activity deletion* sub-model for highly *urgent activities* is provided – this decision would be based on the level of priority associated with A_i' (in most cases, this would result from high priorities being assigned by the *random event simulator*). Otherwise, the role of the MCR is to identify activities and specific modification types that could be made to make room for the new activity. In the case of open time, no A_i' exists, and instead the MCR identifies activities that can be lengthened via a set of modifications to fill the empty time.

In either case, the MCR does this by defining a function similar to (12.1) that determines the 'priority' ($P_j^{m,\text{mod}}$) of scheduled activities A_j for modification (the subscript i is replaced by a j in all cases, and X_j refers to the attributes of activity j on the schedule). Such a model would be estimated using data on observed modifications, such as that provided by the CHASE survey. In particular, variables that indicate the flexibility of each activity for potential modification should be included in the model. For instance, the differences between a scheduled activity's duration (d_j) and the activity's minimum duration (min d_j from the agenda) would be a strong indicator of its potential for modification. The proximity of A_j to the window of opportunity, in terms of the number of activities that separate the two, may also be an important determinant, as closer activities are more likely to satisfy the modification need. The number of locations and possible modes available to the user may also signify more flexibility for modification.

Similar to the addition of activities, once the priority for modification is established, a choice must made of which A_j to modify (*choose* A_j *to modify*), similar to the choice of activities to add in the main model in Figure 12.1 (*choose activity* A_i *to add*). The next step is to *determine the modification types for* A_j and their potential to free up time in order to increase the size of W_n' (in the case of open time, the goal would be the opposite – to fill up time in order to reduce the W'_n). These include just moving the activity in time, reducing the duration in relation to its minimum, and changing the location or mode to reduce travel time. The maximum possible savings should be identified (for example the difference between duration of A_j and its minimum possible duration; the difference between the travel time to current location and the travel time to the nearest perceived alternative location).

Once these modifications and their potential are identified, a decision must be made to *choose another activity to modify*. This choice would be based on the increase in the amount of time savings that has accrued from the preceding step compared to the amount of time necessary. Thus, if the ratio of the potential time savings to the time necessary is below a threshold value, then the procedure should stop, similar in structure to the rule in (12.2). This would halt the search for modifications at a point where no significant gains are being made.

At this point in the sub-model, the range of potential modifications must be evaluated in terms of their ability to meet the minimum time needed or, in the case of open time, fill up the time (*evaluate impact of modifications*). This would constitute a simple rule wherein the sum effect of the modifications must be greater than the minimum time necessary. If not, then more drastic measures may be considered in the form of *activity deletions*. If they are adequate, then the magnitude of the modifications must be refined such that any potential savings in time are distributed fairly amongst the new activity (A'_i) and the modified ones. For example, suppose a new activity requires 30 extra minutes in order to be scheduled at its minimum duration, and that another activity could potentially be shortened by as much as one hour in order to make room for the new activity. The decision is then how the residual 30 minutes of activity time is distributed between the two activities. A simple approach would be to split the amount of time savings equally after both activities are scheduled at their minimum. In this case, 30 minutes of the modified activity is traded to the new activity to get it to a minimum, and the residual 30 minutes is distributed equally to both, providing an additional 15 minutes of duration. A more complex rule could distribute any residual time savings in proportion to the durations of the activities, or in proportion to their priority. The complexity of this decision increases as the number of alternative modifications increases. Further analysis of the observed modifications, and more in-depth analysis of the underlying decision process is necessary before more refined rules can be suggested.

Once the refinements are made, the modifications are implemented (*modify activities*) and control reverts to the main model, where the final refinement choices for the new activity are made given the new time window.

Deletions

In the event that no set of modifications is adequate to resolve a conflict or schedule an urgent activity, then the deletion of an activity is considered. A flow diagram of this procedure was presented in Figure 12.3. The procedure is similar to the MCR in the inputs, the priority model ($P_j^{m,\text{del}}$), and the choice of activity to delete. Following this, a decision then needs to be made about whether the new activity is indeed of much higher priority to warrant deletion of the chosen activity (*delete* A_j*?*). One way to handle this would be to re-evaluate the priority of the chosen activity for addition (that is, $P_j^{m,\text{add}}$) given the new state of the schedule (this would differ from when it was originally made) and compare it directly to the priority of the new activity. If the discrepancy is large enough (that is, $P_i^{m,\text{add}} >> P_j^{m,\text{add}}$), then the deletion should proceed, and control revert back to the main model where the resulting time window is then reassessed (note that further modifications or deletions may follow to accommodate the new activity). If not, then other activities may be considered for possible deletion (*choose next activity to delete?*). In the event that no deletion is feasible, then control reverts back to the MCR and subsequently back to the main model, and the activity is no longer considered for addition. At this point, it can be assumed that a rescheduling response was simply not adequate to accommodate the scheduling of the new activity, or that the activity was simply not important enough to justify changes to existing activities.

Optimization in the Scheduling Process

Although the scheduling process proceeds in a sequential fashion, without directly involving the optimization of the schedule as a whole (apart from the optimization already achieved through scheduling high-priority activities first), a further degree of optimization is achieved in the scheduling model by revisiting previous activities for modification in the event that *time pressure* accrues, or *open time* results. This leads in the first case to more optimized locations, durations, mode choices, and so on that minimize travel time or durations via the *modify and conflict resolver*, and in the second case by allowing previously planned activities to be optimized in a sense that preferred locations or extended durations are realized. Thus, durations and travel times (via location choices) will only be optimized to the extent that other activities of high priority need also be scheduled in the same time window. Behaviourally, this reflects the notion that people consider optimizing their behaviour only when and where needed and/or possible.

5. CONCLUSIONS

This chapter has sought to move closer to the solution of a complex problem in the field of travel behaviour that has hampered development of behaviourally sound forecasting models – namely to understand better how various activity/travel choices or scheduling decisions are interrelated over time. An attempt was made to provide a behaviourally realistic framework which brings in the important dynamic component of household activity/travel scheduling. This development is part of a continuing and long-term project for the authors. One of the main goals is to support the development of an operational scheduling model to be used within integrated land-use, transportation and environmental emissions models currently being developed by a collaborative research team in Canada (Miller and Salvini, 1998).

Empirical evidence obtained by means of a newly developed survey methodology (CHASE; see Doherty and Miller, 2000) was used to support the development of the conceptual framework. Furthermore, the data make possible future development in the form of the specific estimation of several sub-models so that assumptions of the conceptual framework as a whole can be tested and validated. In the case of the momentaneous priority models, for instance, they may be estimated using the existing CHASE survey data, with careful selection of variables from Table 12.2 and due consideration to the form of the model. Some of the decision rules may also be investigated using the CHASE data. However, their exact form would benefit from further parallel investigation of underlying cognitive decision-making processes.

The conceptual framework also places renewed focus on basic human needs, as manifested in the household activity agenda. Simulation of a detailed activity agenda is crucial to the scheduling model, and represents a significant research challenge. Such development may benefit from contemporary developments in motivational psychology (Gärling and Garvill, 1993). The importance of the agenda is also highlighted by the fact that many changes in policy would first need to be implemented via changes in the attributes or distribution of activities on the agenda. This would then invoke a scheduling response and subsequent changes in activity and travel patterns. For instance, the implementation of flexible work hours would in the first instance lead to changes in the earliest start and latest end time for work activities on the agenda, which would then have clear implications for scheduling within the household.

The conceptual model also leaves room for a role to be played by existing optimization style models. It attempts to address past behavioural criticisms of these models, by providing a reduced choice set of activities that are amenable to optimization. Applying past successes with these models to a weekly scheduling setting is an important aspect of the success of the overall scheduling model.

Overall, the CHASE survey and results went a long way towards solving the data collection problems that have hampered the development of simulation models of activity scheduling, of which the current one is but one possibility. Continued focus on activity scheduling, with the support of a new technique such as CHASE, could lead to fundamentally new advances in the future.

ACKNOWLEDGEMENTS

This study was financially supported in part by a doctoral fellowship awarded to Sean Doherty by the Social Science and Humanities Research Council of Canada, a collaborative research grant to Eric Miller from the Natural Science and Engineering Research Council of Canada, and a grant to Tommy Gärling from the Swedish Transport and Communications Research Board.

REFERENCES

Axhausen, K.W. and T. Gärling (1992), 'Activity-based Approaches to Travel Analysis: Conceptual Frameworks, Models, and Research Problems', *Transport Reviews*, **12**(4), 323–41.

Bowman, J.L. and M. Ben-Akiva (1997), 'Activity-based Travel Forecasting', Summary, *Recommendations and Compendium of Papers from the Activity-based Travel Forecasting Conference*, New Orleans, pp. 3–36.

Cullen, I. and V. Godson (1975), 'Urban Networks: The Structure of Activity Patterns', *Progress in Planning*, **4**(1), 1–96.

Doherty, S.T. and E.J. Miller (2000), 'A Computerized Household Activity Scheduling Survey', *Transportation*, **27**(1), 75–97.

Ettema, D., A. Borgers and H. Timmermans (1993), 'Simulation Model of Activity Scheduling Behaviour', *Transportation Research Record*, **1413**, 1–11.

Ettema, D., A. Borgers and H. Timmermans (1994), 'Using Interactive Computer Experiments for Identifying Activity Scheduling Heuristics', paper presented at The Seventh International Conference on Travel Behaviour, Santiago, Chile.

Ettema, D. and H. Timmermans (eds) (1997), *Theories and Models of Activity Patterns. Activity-based Approaches to Travel Analysis*, Oxford: Pergamon, pp. 1–36.

Gärling, T. (1998), 'Behavioural Assumptions Overlooked in Travel Choice Modeling', in J. Ortuzar, S. Jara-Diaz and D. Hensher (eds), *Transport Modeling*, Oxford: Pergamon.

Gärling, T., K. Brannas, J. Garvill, R.G. Golledge, S. Gopal, E. Holm and E. Lindberg (1989), 'Household Activity Scheduling', *Transport Policy, Management & Technology Towards 2001*, Ventura: Western Periodicals, pp. 235–48.

Gärling, T. and J. Garvill (1993), 'Psychological Explanations of Participation in Everyday Activities', in T. Gärling and R.G. Golledge (eds), *Behavior and Envi-*

ronment: Psychological and Geographical Approaches, Amsterdam: Elsevier/North-Holland, pp. 270–97.

Gärling, T., T. Kalèn, J. Romanus, M. Selart and B. Vilhelmsson (1998), 'Computer Simulation of Household Activity Scheduling', *Environment and Planning*, **30**A, 665–79.

Gärling, T., M. Kwan and R. Golledge (1994), 'Computational-process Modelling of Household Activity Scheduling', *Transportation Research*, **28**B, 355–64.

Hayes-Roth, B. and F. Hayes-Roth (1979), 'A Cognitive Model of Planning', *Cognitive Science*, **3**, 275–310.

Jones, P., F. Koppelman and J.P. Orfeuil (1990), 'Activity Analysis: State-of-the-art and Future Directions', in P. Jones (ed.), *Developments in Dynamic and Activity-based Approaches to Travel Analysis*, Aldershot: Avebury, pp. 34–55.

Kitamura, R. (1983), 'Sequential, History-dependent Approach to Trip-chaining Behavior', *Transportation Research Record*, **944**, 13–22.

Kitamura, R. and S. Fujii (1998), 'Two Computational Process Models of Activity-travel Choice', in T. Gärling, T. Laitila and K. Westin (eds), Amsterdam: Elsevier, pp. 251–79.

Miller, E.J. and P.A. Salvini (1998), 'A Microsimulation Approach to the Integrated Modeling of Land Use, Transportation and Environmental Impacts', paper presented at the 76th Annual Meeting of the Transportation Research Board, Washington, DC.

Pas, E.I. (1985), 'State-of-the-art and Research Opportunities in Travel Demand: Another Perspective', *Transportation Research*, **19**A, 460–64.

Payne, J.W., J.R. Bettman and E.J. Johnson (1993), *The Adaptive Decision Maker*, New York: Cambridge University Press.

RDC Inc. (1995), *Activity-based Modeling System for Travel Demand Forecasting*, Metropolitan Washington Council of Governments Washington, DC:US Department of Transportation.

Recker, W.W. (1995), 'The Household Activity Pattern Problem: General Formulation and Solution', *Transportation Research*, **29**B(1), 61–77.

13. Effects of office relocations to public transport nodal points on passenger mobility

Bert Van Wee and Toon Van Der Hoorn

1. INTRODUCTION[1]

Dutch location policy attempts to achieve labour-intensive employment (large number of employees per unit area) situated close to public transport nodal points. More than on existing employment locations, attention is focused on new situations. The consideration of firm relocation in this policy has been limited until now. In the Netherlands, however, about 6–8 per cent of firms move every year. In 1992 the absolute number of firms moving was 58 000, involving 180 000 employees (Pellenbarg, 1996). At least 70 per cent of all moves are over a short distance (within the same town or conurbation) (Van Steen and Van Der Velde, 1993). The question of employees' reactions is therefore very relevant. Several years ago, the National Institute of Public Health and the Environment (RIVM) started a research programme to establish the possible effects on mobility of a large-scale relocation of existing office employment to sites easily accessed by public transport. Following literature surveys, empirical research was carried out. Employees of two relocated offices were surveyed. Models which describe employees' reactions to office relocation (change of dwelling yes/no, change of job yes/no) have been estimated. These models have been applied in a scenario study to evaluate the effects of a large-scale office relocation for a period up to 2015 to sites easily accessed by public transport. Van Wee (1997) gives a summary of the complete research programme. This chapter focuses on the spatial scenario study, and includes summaries of the Rijkswaterstaat survey and of the models used. Furthermore, the following methodological issue will be dealt with: do state-of-the-art models estimate effects of land-use scenarios well?

Section 2 outlines Dutch location policy. Section 3 briefly presents the empirical survey. Section 4 presents the model estimates for employees' reactions to office moves. Section 5 deals with the set-up of the scenario study, with section 6 describing how the new models were combined with a more or less traditional transport model. Section 7 gives the main results from the scenario study and section 8 deals with the relevance of the results for Dutch location policy. Section 9 answers the question: do state-of-the-art models estimate effects of land-use scenarios well? Section 10 gives final conclusions.

2. AN OUTLINE OF DUTCH EMPLOYMENT LOCATION POLICY

Central to Dutch policy are the attributes of locations and of businesses. Two concepts are introduced: the 'accessibility profile' of a location and the 'mobility profile' of a firm (VROM et al., 1990 and Van Wee and Van der Hoorn, 1996).

The 'accessibility profile' of a location describes the accessibility by public transport and car. The parking facilities are also part of the accessibility profile. There are three location types:

- Location type A: These are situated close to public transport interchanges of national or regional importance. In cities and towns with an intercity/Eurocity railway station, the area near the station is an A location. Employment density is high. There are few parking facilities. For new offices a maximum of 10 spaces per 100 employees is allowed in the Randstad,[2] and 20 elsewhere. Connections to the motorway system are not of prime importance.
- Location type B: These are situated close to public transport connections of local or regional importance, and near a major local road or motorway connection. Employment density is lower than for A locations. There are also more parking facilities. For new offices a maximum of 20 spaces per 100 employees is allowed in the Randstad and 40 elsewhere.
- Location type C: These are situated close to a motorway connection, in or on the periphery of urban areas. Public transport connections are of no importance and there is no upper limit to the parking capacity being provided.

If necessary, by means of investment in rail (public transport) or road infrastructure, new A, B and C locations can be created and the profile of existing A, B and C locations can be changed. Not all locations satisfy the ABC attributes. On these 'R locations' (remaining locations) further development of the employment base is considered undesirable.

The 'mobility profile' of a firm or public service is defined as the potential use of public transport by employees and visitors. The main characteristics are:

- Labour intensity: the number of employees per unit area
- Car dependence of the firm in performing its business
- Visitor intensity: the number of visitors per unit area
- Dependence on freight transport by road

The essence of the employment location policy is to match accessibility profiles of locations and mobility profiles of companies. The policy applies primarily to new situations and not so much to existing employment locations. The employment location policy is intended for achievement at the regional level. The character of the accessibility profiles will differ by region or agglomeration. Cooperation between national government, provinces and municipalities is necessary. For this, the concept of 'transport regions' has been developed.

3. THE EMPIRICAL SURVEY

Design

The survey for the empirical research was carried out among employees of two government offices that moved about 20 kilometres to one central building in Rotterdam. The new location is situated downtown, is very accessible to public transport and has very little provision for car parking. The survey took place among all employees who worked in the department six months before the relocation, including people who had resigned since then. This means that anticipation behaviours could be observed as well. It is known from the literature that people may already move or change job before the actual firm relocation takes place.

The survey was carried out 4.5 years after the office relocation. The panel data set obtained includes variables such as changes in dwelling and working locations and in personal and household data over a period of 5 years. Because collection of continuous panel data over a period of 4.5 years would have been awkward, retrospective questioning has been used instead. The advantage is that data collection is relatively cheap and that data are readily available. The disadvantage is the chance of a selective non-response: employees who have changed employer as well as dwelling are difficult to trace. In this study, it was possible to obtain practically all the addresses of 'lost' employees via former colleagues. Another possible disadvantage may be bias, as a result of the fact that people neither perfectly remember what they did or did not do, nor their reasons, from some years ago. We think this disadvantage does not occur in our research because the questions are only related to major decisions (change/no change of job/residential location), major characteristics of homes (buy/rent, type) and travel (mode, travel time), and major (changes in) personal and household characteristics (household composition, income, education). If (unnoticed) bias exists, the influence on applications of the model will be limited in view of the clear, general, conclusions (see below).

Network data for car and public transport were added to the survey data. These concern travel distances and travel times between various living and working locations before and after the office relocation. Using these, it can, for example, be evaluated whether a short commuting distance (before the office relocation) increases the probability of moving with the employer.

Results

More than half (58 per cent) of the respondents have not moved and have stayed with the same employer. About one-quarter (23 per cent) have moved, but have stayed with the same employer. While 8 per cent have changed employment, they have not moved. Six per cent have both changed employment and dwelling and 4 per cent have stopped working; of these, only one person has moved. Change of employment is often accompanied by a change of dwelling, but normally not as a consequence of an office relocation.

Many respondents have moved within the same residential area. Only a small number have moved towards the new office location. Therefore, even after 4.5 years, the relocated offices are still situated ex-centric to the dwellings of employees. Employees who have changed jobs have on average about the same commuting distance as before the change.

Spatial behaviour (change of dwelling yes/no, change of job yes/no) is related to a large number of *personal and household* variables (such as income, age and household size), to various characteristics of the *transport system* (such as the commuting distance before the office relocation and the degree to which employees' commuting distance would increase due to the office relocation without changing residential location), and to *interactions* between variables.

4. MODELS FOR EMPLOYEES' REACTIONS TO OFFICE RELOCATION

To evaluate the effects of a relocation of employment, one can use a 'traditional' transport model, such as the National Model (LMS) of the Ministry of

Transport (Bovy et al., 1992). Because such a model is usually based on cross-sectional data, it calculates the long-term equilibrium situation between demographic, economic and spatial developments. To account for the medium-term effects (about five years after an office relocation), employees' reactions have been modelled using the survey database. The resulting models are intended to be embedded within and to enhance traditional models. Two types of models have been estimated:

- A model which describes the reaction type of employees when confronted with an office relocation (change of dwelling yes/no, change of job yes/no)
- A model which describes whether employees changing dwelling and not changing job will move within the same residential area (because of life-cycle reasons) or to another residential area in the direction of the new job location.

The models contain person and household variables and level-of-service variables. The degree to which employees' commuting distance would increase due to the office relocation in the event that they do not change residential location appears to be especially important. The model describing the reaction of employees when confronted with an office relocation (change of dwelling yes/no, change of job yes/no) is given in Appendix 1, Table 13A.1. Table 13A.2 provides the model that describes whether a worker who changes dwelling, but not employer, will move within the *same* residential area or to *another* residential area, in the direction of the new work location. Only employees who would be disadvantaged by a distance of at least 14 km (when not moving and changing jobs) are considered to have a choice of moving to another residential area. In other words, moving employees who are less than 14 km from their office move by definition (in the model) within the same residential area.

For those employees moving to another residential area, in the direction of the new office location, a new dwelling zone has to be established. This has been done using an algorithm deduced from the empirical survey as there were not enough observations to estimate a model. The number of variables in the models is limited for two reasons. First, the number of cases that are available in the data set is limited (217), making it impossible to draw conclusions with respect to some possibly relevant variables, for example only 8 per cent of the respondents are females. Second, not all variables in the data set are available in the LMS. Although these variables were initially included in the model estimation (Van Wee, 1996a), they were excluded in the models as applied to the scenario study as described in this chapter. We feel that the fact that the models contain only a limited number of variables does not influence the simulation results because the general conclusions are very clear: a vast majority of the respondents do not change job and do not move, or move only a short distance.

5. A SPATIAL SCENARIO FOR RELOCATION OF OFFICE EMPLOYMENT

A scenario for 2015 was evaluated, in which, in essence, 50 per cent of Dutch employment at C and R locations is transferred to A and B locations. The basis for the expected socio-economic development pattern is the European Renaissance (ER) scenario of the Netherlands Bureau for Economic Policy Analysis (CPB, 1992). Furthermore, 'current' policy is assumed, for example with respect to pricing and infrastructure policy and emission standards for road vehicles. The combined socio-economic and policy scenario was developed to show the effects of 'current policy' and so provide the basis for the long-term strategic environmental policy as presented in the National Environmental Policy Plan 2 of 1993 (see Van Wee, 1996b). To show explicitly the effects of the policy, this policy is considered exogenous, and not endogenous, although in practice (environmental) trends may influence government policy. Of course, not all employment is eligible for transfer. Therefore retail and agricultural employment are excluded. Also, particular rules have been defined to select plausible A and B locations as destinations for the transfer. The maximum distance over which transfer of firms takes place is 20 km. The resulting scenario has been submitted for comment to the Spatial Planning Agency (RPD) of the Dutch Ministry of Housing and the Environment (VROM), to assess whether 'impossibly high' concentrations of employment could arise. This appeared not to be the case, although higher land costs could result. The total number of transferred jobs is 796 000, or 12 per cent of all jobs in the ER scenario. The scenario is suitable for rough planning purposes, but not for detailed forecasting and policy evaluation, having insufficiently fine detail at the regional level. Because the choice models were estimated according to a survey giving the reactions of employees 4.5 years after the office relocation, it is implicitly assumed that the firms are transferred in (a suitable interval around) 2010.

6. METHODOLOGY: THE COMBINED USE OF A TRADITIONAL MODEL WITH THE MODELS FOR EMPLOYEES' REACTIONS

To calculate the effects of the spatial scenario, the National Model System (LMS) was used. It is a somewhat traditional transport model, containing state-of-the-art disaggregate models (see Bovy et al., 1992). The models for the reactions of employees to office relocations were combined with the LMS. First, forecasts are made using the standard LMS for the reference situation 2015, without relocation of employment. In this step, reference Origin Destination (O–D) matrices for the purpose of 'commuting' are derived. Hence, for each zone, the origins (dwellings) and commuting times, by mode, of people working in that zone are known. Next, job locations are transferred according to the spatial scenario chosen. For the job locations transferred, the spatial behaviour of employees is evaluated using the explanatory variables (including level-of-service variables) and the models developed in the study. Once new dwellings and work locations are known, new O–D matrices can be derived. The rules for assignment of new dwelling zones and employment zones to people with various reaction types are given in Table 13.1.

In the case of a *new dwelling* location (reaction type 3(B)), the procedure is as follows. From the Rijkswaterstaat survey, it appears that the new dwelling location of employees who have moved is on average about 7.5 km from the new job location. Hence the dwelling locations are assigned to the LMS sub-

Table 13.1 Assignment in the scenario of new dwelling and job location zones after office transfer

	Reaction type	Dwelling assignment	Job location assignment
1	Not moved, same employer	New dwelling location = old dwelling location	Take as new commuting destination the zone where the office has been transferred
2	Not moved, other employer	New dwelling location = old dwelling location	New job location = old job location[a]
3	Moved, same employer		
3(A)	Small reduction of commuting distance (10 km or less)	New dwelling location = old dwelling location	Take as new commuting destination the zone where the office has been transferred to
3(B)	Large reduction of commuting distance (10 km or more)	Determine new dwelling location as described	Take as new commuting destination the zone where the office has been transferred to
4	Move, other employer	New dwelling location = old dwelling location	New job location = old job location

Note: [a] The commuting distance in the new situation is practically the same as that in the old situation (see Van Wee, 1995, pp. 276–7).

zones surrounding the work zone so that the average distance (weighted by number of houses) to the work zone is 7.5 km. To keep the total number of households constant, some existing households have been 'removed' from those zones where other households have reassigned. Likewise, 'new' households have been 'added' to those zones from which households have been removed. These 'added' households are assumed to have the same characteristics (employment location, commuting travel mode, and so on) as the households already located in the zone.

The ABC parking restrictions in the Dutch location policy are also assumed to be applicable to the transferred job locations. This means that the number of parking places increases proportionally to the number of jobs. The modal split has been left unchanged for every O–D relation, although in reality (marginal) changes might occur due to changed congestion levels on trunk roads.[3] The choice of an alternative parking location is considered a separate travel mode.

Various personal and household data are not available in the computer when the spatial choice models for employees' reactions to office relocation are in operation. It is, therefore, assumed that the labour force is distributed uniformly over the Netherlands. In other words, employees of transferred firms in, for example, Amsterdam have the same distribution over such items as age, income and education as employees of transferred firms, for example in Groningen (in the north of the country).

As explained before, the spatial choice models for employees' reactions to office relocation describe behaviour in the medium term, because they take lags of response behaviour into account. The results can be compared with the normal results from the standard application of the LMS for the long term. In this case, standard application of the LMS means that job locations are transferred to sites easily accessible to public transport. This causes the zonal attraction variables in LMS to change. The next step is the application of the standard (cross-sectional) suite of the tour frequency and distribution/mode split sub-models.

7. RESULTS

The assumed office relocation especially affects commuting travel, although other travel purposes are also affected. In this chapter, only results for commuting travel are given (Table 13.2). From the table, it appears that the spatial scenario increases total kilometrage (all travel modes) in the long-term equilibrium situation by 2.4 per cent. The calculated increase over the medium term using the models for employees' reactions to office relocation is 3.2 per cent. This is logical: the pattern of living and working locations will adapt

Table 13.2 Kilometrage commuting traffic per scenario (2015)

Travel mode	Index unchanged policy	Transfer of employment: LT effects (index w.r.t. unchanged policy)	Transfer of employment: MT effects (index w.r.t. unchanged policy)
Car driver	100	96.0	98.7
Car passenger	100	116.6	117.0
Public transport	100	117.2	113.2
Slow	100	104.5	101.8
Total	100	102.4	103.2

Notes:
Unchanged policy: European Renaissance scenario, current government policy.
Transfer of employment: LT effects: transfer of employment, long-term equilibrium situation as calculated by LMS.
Transfer of employment: MT effects: transfer of employment, medium-term situation, as calculated by the spatial choice models for employees' reactions to office relocation developed in this study.

itself only in the long term. In the medium term, most employees do not move with their employer. Therefore the firms transferred are situated in the first instance (that is, in the medium term) ex-centric with respect to their employees. Furthermore, it appears that the kilometrage of the car driver is reduced both in the medium and in the long term, and that the kilometrage by other modes is increased. The increase for the slow modes is smaller than that for car passengers and for public transport.

From the table, it is also clear that the effects in the medium term are markedly different from those in the long term. For example, the reduction in car kilometrage is 4.0 per cent in the long term, but only 1.3 per cent in the medium term, using the models for employees' reactions to office relocation. Likewise, the increase of the slow modes in the long term (4.5 per cent) is much larger than that in the medium term (1.8 per cent).

The effects of the spatial scenario, at first glance, seem limited. The effects, however, are expressed relative to the total journey to work kilometrage in the Netherlands. Therefore it can be inferred that by transferring (only) 12 per cent of employment, car kilometrage is reduced by 1.3 per cent (in the medium term). In the long term the reduction is (even) 4.0 per cent. The share of the transferred employment in total car kilometrage is 13 per cent and therefore a bit higher than the share in total employment of 12 per cent. This is because mainly firms at C and R locations are involved, which have a relatively high share of car use. The calculated reduction of car use in commuting of employees concerned is 10 per cent (medium term) to 31 per cent (long term).

8. RELEVANCE FOR LOCATION POLICY

The results of the spatial scenario raise the question of whether policies aimed at the transfer of jobs to locations easily accessed by public transport should take account of lags in the reactions of employees. These lags have different effects in the medium and long term: often the effects are larger in the long term. The effects of spatial policy may take years to materialize. If we focus exclusively on the long term, we may overestimate the effects of spatial policy. It is even questionable whether the long-term equilibrium will ever be reached. For many spatial and infrastructural measures, continuous disequilibrium may more often be the rule than long-term equilibrium. Given the dynamics in the office sector, for example, firms might already have moved once again before the long-term equilibrium has been reached.

What recommendations can be made from the results of the study? The general picture is that transfer of offices to locations easily accessed by public transport may cause substantial reduction of car use and substantial increase in public transport use, certainly in the long term. We believe, however, that no policy of 'force' should be used. The costs of transferring firms may be very high and may not be in proportion to the benefits of accessibility and the environment. It may be more productive to find firms which have already planned to move (see Heidemij et al., 1991). Likewise, to firms situated at C and R locations the advantages of a railway location could be presented. Once these firms are prepared to move, 'tailor-made' measures can be designed to stimulate them to relocate to A or B locations. The (future) availability of capacity (land, office buildings) is a necessary condition for such a spatial policy. Also, improvements in public transport and bicycle infrastructure, and possibly subsidies, will play a role. Finally, the quality of the railway station environment is important. This can be improved by preventing mono-functionality (work only) and by situating amenities (especially shops and restaurants) and housing near railway stations. Liveliness and objective and subjective safety would be improved and the contrast with the neighbouring historical city centre reduced. One may think of buildings where the lower floors are used for shops and restaurants, the middle floors for apartments, and the top floors for offices. We think that discussion between government and the private sector, and an information system for requirements of and potential for successful relocation are more important than just a pot of 'money'. In some instances, however, money can act as a motivator.

9. DO STATE-OF-THE-ART MODELS ESTIMATE EFFECTS OF LAND-USE SCENARIOS WELL?

For five reasons we think that state-of-the art models overestimate effects of land-use scenarios, mainly expressed as kilometres per mode.

First, traditional models do not use fixed time budgets. Therefore total travel time may vary between land-use scenarios. However, from several studies we know that on an aggregate level people have a more or less constant travel time budget (see Kraan, 1996, for an overview). Therefore traditional models may overestimate the variation in mobility between land-use (and other) scenarios.

Second, state-of-the-art models do distinguish between household or personal classes, but within a single class people are assumed to be homogeneous. Mobility differences between land-use and transport scenarios are mainly the result of differences in generalized travel costs (time, money) between modes. However, it is possible that there is a spatial bias: especially in the case of a coarse zoning system, people within household classes may vary with respect to their preferences of travel mode and therefore with respect to their residential location. People who prefer to travel by train may prefer a residential location close to a railway station. People without a car do not consider a residential location that is inaccessible by public transport (Pickup and Town, 1983). This spatial bias might lead to an overestimation of variation in mobility between land-use and transport scenarios.

Third, traditional models, calculating the long-term equilibrium, do not consider the incremental character of changes. The time of changes may influence long-term effects. For example, if public transport is implemented in new residential areas long after the first residents have moved in, fewer people may travel by public transport than if public transport is already available right from the beginning. Travel behaviour has a strong habitual component (Van Vugt, 1996). Feedbacks as assumed in systems dynamics models do not play a role in traditional models. Panel-data-based models sometimes show that effects of changes may be smaller than cross-section-based models predict (Meurs, 1991).

Fourth, traditional models assume that a long-term equilibrium exists. It can be argued whether this long-term equilibrium will ever be reached, especially in the case of land-use and transport scenarios that differ strongly from the current situation or the trend. For instance, large differences in employment data per zone between the base year and the forecasting year implicitly assume firm relocations. As shown in this study, the situation five years after an office relocation differs strongly from the long-term equilibrium. Most people do not move and do not change jobs. Therefore average home–work distances are longer than according to the long-term equilibrium.

Fifth, part of total travel is intra-zonal and thus over short distances. Short-distance travel is less sensitive to measures than other travel. Especially if zones are relatively large, models may overestimate effects of measures/changes.

For all these reasons, we think that the direction shown in differences between land-use scenarios as calculated with traditional models is right but that the differences are overestimated. Apart from these reasons there are two other factors that may be important. For these factors, however, we do not know a priori whether differences will be over- or underestimated.

1. The interaction between transport and land use. This interaction is not incorporated in traditional models. These interactions may lead both to an overestimation and an underestimation of differences in mobility between land-use scenarios.
2. Mobility chains are not normally included in traditional models. People sometimes combine trips. From time–space literature we know that the home–work axis strongly influences total mobility. Land-use scenarios very often differ with respect to residential and working locations. The influence of such land-use scenarios on total travel may be larger than calculated.

10. CONCLUSIONS

Dutch location policy tries to get 'the right firm in the right location'. But the focus on office relocation has been limited until now. Every year, however, about 6–8 per cent of firms move. The number of jobs involved in office relocation every year is relatively large (1992: about 180 000). A large majority of these moves takes place over a short distance (about 70 per cent within the same conurbation). More attention to office moves is therefore desirable.

In the case of an office relocation of about 20 km or less the overwhelming majority of employees do not change employer and do not move or only move within the same residential area. Therefore in evaluating the effects of employment relocation, one has to estimate explicitly the (delayed) effects on employee behaviour in the medium term to be able to evaluate the full effects on mobility. It is not sufficient to deal with the 'long-term equilibrium situation'. If one focuses exclusively on the long term, one may overestimate the effects of spatial policy. It is even questionable whether the long-term equilibrium will ever be reached. For many spatial and infrastructural measures, continuous disequilibrium may more often be the rule than long-term equilibrium. Given the dynamics in the office sector, for example, firms might already have moved once again before long-term equilibrium has been reached.

Using a limited number of cases, it has been shown to be feasible to develop and estimate spatial choice models for employees' reactions to office relocation over distances of up to 20 km (change of dwelling yes/no, change of job yes/no). The models contain personal and household variables, and level-of-service variables. Especially important appears to be the degree to which employees' commuting distance would increase due to the office relocation if they did not change residential location.

When panel data are desirable, but when it is impossible to interview households on several occasions, retrospective questioning can be used instead. The advantage is that data collection is relatively cheap and that data are readily available. A necessary condition, however, is a good record of households in the pre-study situation and a good tracking system for categories of households which have a relatively high probability of non-response.

The estimated models have been fairly easily incorporated into the standard Dutch National Model System (LMS). The models have been applied in a scenario study to evaluate the effects of a large-scale office relocation for a period up to 2015 to sites easily accessed by public transport. The results are highly dependent on whether the delayed reactions of employees to office relocation are explicitly dealt with. The effects on car kilometres, for example, are three times as large in the long-term equilibrium situation than in the medium-term situation once the delayed reactions are taken into account.

A scenario for 2015 was evaluated, in which in essence 50 per cent of Dutch employment at C and R locations is transferred to A and B locations, easily accessed by public transport. Car kilometres of affected employees are reduced by 30 per cent in the long term, but only by 10 per cent in the medium-term situation, when delayed reactions are taken into account.

State-of-the art models overestimate effects of land-use scenarios.

NOTES

1. This chapter is a revised version of the paper by Van Wee et al. (1997).
2. The Randstad is the highly urbanized western part of the Netherlands, including the for largest cities.
3. Hence the LMS sub-models themselves are not used in the evaluation of the land-use policy measures in the scenario. These sub-models are only used to derive the reference O–D matrices for the base situation without land-use policy.

REFERENCES

Bovy, P., J. Jager and H. Gunn (1992). 'The Dutch National and Regional Model Systems: Principles, Scope and Applications', *Selected Proceedings of the Sixth World Conference on Transport Research (WCTR)*, **2**: Demand, Traffic and Network Modeling. WCTR Society, pp. 1197–208.

Central Planning Bureau (CPB) (1992), *Scanning the Future. A Long-Term Scenario Study of the World Economy 1990–2015*, The Hague: SDU (Government Printing Office).

Heidemij, L.U.M., T.U. and Seinpost (1991), *Stedelijke Herverkaveling en Mobiliteitsprofielen van Bedrijven* (Urban Relocation and Mobility Profiles of Firms), Heidemij: Arnhem.

Kraan, M. (1996), *Time to Travel? A Model for the Allocation of Time and Money*, Enschede: Universiteit Twente.

Meurs, H.J. (1991), *A Panel Data Analysis of Travel Demand*, Groningen: Rijksuniversiteit Groningen.

Pellenbarg, P. (1996), 'Structuur en Ontwikkeling van Bedrijfsmigratie in Nederland: Bedrijfsmigratie als Signaal van Ruimtelijke Ontwikkelingstendenzen' (Structure and Development of Office Migration in the Netherlands: Office Migration as Signal of Spatial Development Trends), *Planning Methodiek & Toepassing*, 32–42.

Pickup, L. and S.W. Town (1983), 'Commuting Patterns in Europe: An Overview of the Literature', TRRL Supplementary Report 796, Berkshire: Transport and Road Research Laboratory.

Van Steen, P. and B. Van der Velde (1993), *Expansieruimte voor Bedrijven. Mogelijkheden en Onmogelijkheden voor uitbreiding van bedrijf spanden* (Space for expansion of firms. Possibilities and impossibilities for expansion of buildings for firms) Groningen: Faculteit der Ruimtelijke Wetenschappen Rijksuniversiteit Groningen. Onderzoek en Advises 69.

Van Vugt, M. (1996), *Social Dilemmas and Transportation Decisions*, Southampton: University of Southampton Design and Print Centre.

Van Wee, G.P. (1995), *Ruimtelijke en Mobiliteitsreacties van Werkenden op Bedrijfsverplaatsingen. Resultaten van Empirisch Onderzoek* (Spatial and Mobility Reactions of Employees to Office Relocation. Results of Empirical Research), Report no. 408134001, Bilthoven: Rijksinstituut voor Volksgezondheid en Milieu.

Van Wee, G.P. (1996a), *Modellen voor ruimtelijke reacties van werkenden op bedrijfsverplaatsingen* (Models for Spatial Reactions of Employees on Firm Relocations), Report no. 715651001, Bilthoven: Rijksinstituut voor Volksgezondheid en Milieu.

Van Wee, B. (1996b), 'The Treatment of Traffic in the Dutch Environmental Outlooks: The Role of Science in Policy Making and Policy Evaluation', *Transportation Planning and Technology*, **19**, 265–74.

Van Wee, G.P. (1997), *Kantoor naar het Spoor. De Invloed van Bedrijfsverplaatsingen naar Openbaar-Vervoerknooppunten op de Personenmobiliteit* (Office of the Railways: The Influence of Office Relocation to Public Transport Nodal Points on Passenger Mobility), Bilthoven: Rijksinstituut voor Volksgezondheid en Milieu.

Van Wee, G.P. and A. Van der Hoorn (1996), 'Employment Location Policy as an Instrument of Transport Policy in the Netherlands. Fundamentals, Instruments and Effectiveness', *Transport Policy*, **3**(3), 81–9.

Van Wee, B., T.D. Bakker and T. Van Der Hoorn (1997), 'Office Suites Suit the Railways: The Effects of Office Locations to Public Transport Nodal Points on

Passenger Transport', *PTRC European Transport Forum, Annual Meeting, Proceedings of Seminar E: Transportation Planning Methods*, **1**, 269–82.
VROM, V & W, and E2 (1990), Geleiding van de mobiliteit door een locatiebeleid voor bedrijven en voozieningen, Workdocument (Guiding mobility by a location policy for firms and services, working paper), The Hague: Rijksplanologische dienst.

APPENDIX 1

Table 13A.1 Model for reaction type

Choice alternative	Variable	Class	Value of parameter	SD	t–value
1 Not moved; same employer					
1	constant		3.352	0.758	4.4
1	age	≤ 45	–2.275	0.634	–3.6
1	education	≤ junior high school	1.905	1.07	1.8
2 Not moved; other employer					
2	constant		–0.139	0.536	–0.3
	income	< NLG 37 000	–1.111	0.655	–1.7
3 Moved; same employer					
3	constant		0.126	0.539	0.2
	income	NLG 37 000–51 000	1.020	0.355	2.9
3	household situation	single, living with parents	1.122	0.389	2.9
3	education	≤ junior high school	1.733	1.11	1.6
3	difference in car distance	advantage; or less than 10 km disadvantage	–0.969	0.376	–2.6
4 Moved; other employer					
4	income	NLG 37 000–51 000	0.458	0.568	0.8
4	education	senior high school	–2.140	1.06	–2.0
4	difference in car travel time	<10 min	–0.986	0.574	–1.7

Notes:

Number of cases:		217
Number of times chosen:		
	alternative 1:	129
	alternative 2:	20
	alternative 3:	53
	alternative 4:	15
Initial log likelihood:		–300.8 2
Same with constant:		–229.5 6
Final log likelihood:		–198.3 0
Rho square w.r.t. zero coefficients:		0.34
Same w.r.t. model with just constants:		0.14

Table 13A.2 Model for the probability of moving to another residential area for employees who change dwelling but not job

Variable	Class	Value of parameter	SD	t-value
Commuting distance	0–8 km (car)	0.451	1.11	0.4
Interaction term[a]	1	–2.357	1.02	–2.3
Constant		–0.273	0.924	–0.3
Probability 0 coefficients			–21.48	
Probability only constants			–16.55	
Final probability			–13.45	
Rho square w.r.t. 0 coeff.			0.374	
Rho square w.r.t constants			0.187	

Note: [a] The interaction term is either 0 or 1. If net income is NLG 37 000–51 000 and education level is at least HBO (higher education/polytechnic), then its value is 0; in all other cases it is 1.

Index

CARDIFF UNIVERSITY * PRIFYSGOL CAERDYDD *

161 099442 4